LUBRICATING GREASE
MANUFACTURING TECHNOLOGY

LUBRICATING GREASE
MANUFACTURING TECHNOLOGY

Yu. L. Ishchuk

PUBLISHING FOR ONE WORLD

NEW AGE INTERNATIONAL (P) LIMITED, PUBLISHERS
New Delhi • Bangalore • Chennai • Cochin • Guwahati • Hyderabad
Jalandhar • Kolkata • Lucknow • Mumbai • Ranchi
Visit us at www.newagepublishers.com

Copyright © 2005, New Age International (P) Ltd., Publishers
Published by New Age International (P) Ltd., Publishers
First Edition: 2005
Reprint: 2011

All rights reserved.

No part of this book may be reproduced in any form, by photostat, microfilm, xerography, or any other means, or incorporated into any information retrieval system, electronic or mechanical, without the written permission of the copyright owner.

Branches:

- 37/10, 8th Cross (Near Hanuman Temple), Azad Nagar, Chamrajpet, **Bangalore**-560 018.
  Tel.: (080) 26756823, Telefax: 26756820, E-mail: bangalore@newagepublishers.com
- 26, Damodaran Street, T. Nagar, **Chennai**-600 017. Tel.: (044) 24353401, Telefax: 24351463
  E-mail: chennai@newagepublishers.com
- CC-39/1016, Carrier Station Road, Ernakulam South, **Cochin**-682 016. Tel.: (0484) 2377004, Telefax: 4051303.
  E-mail: cochin@newagepublishers.com
- Hemsen Complex, Mohd. Shah Road, Paltan Bazar, Near Starline Hotel, **Guwahati**-781 008. Tel.: (0361) 2513881.
  Telefax: 2543669, E-mail: guwahati@newagepublishers.com
- 105, 1st Floor, Madhiray Kaveri Tower, 3-2-19, Azam Jahi Road, Nimboliadda,
  Hyderabad-500 027. Tel.: (040) 24652456, Telefax: 24652457,
  E-mail:hyderabad@newagepublishers.com
- RDB Chambers (Formerly Lotus Cinema)106A, 1st Floor, S.N. Banerjee Road,
  Kolkata-700 014. Tel.: (033) 22273773, Telefax: 22275247,
  E-mail:kolkata@newagepublishers.com
- 16-A, Jopling Road, **Lucknow**-226 001. Tel.: (0522) 2209578, 4045297, Telefax: 2204098
  E-mail: lucknow@newagepublishers.com
- 142C, Victor House, Ground Floor, N.M. Joshi Marg, Lower Parel, **Mumbai**-400 013.
  Tel.: (022) 24927869. Telefax: 24915415, E-mail: mumbai@newagepublishers.com
- 22, Golden House, Daryaganj, **New Delhi**-110 002. Tel.: (011) 23262370, 23262368,
  Telefax: 43551305. E-mail: sales@newagepublishers.com

ISBN: 978-81-224-1668-8

Rs. 300.00

C-10-06-4665

Printed in India at Glorious Printers, Delhi.
Typeset at Goswami Associates, Delhi.

PUBLISHING FOR ONE WORLD
NEW AGE INTERNATIONAL (P) LIMITED, PUBLISHERS
4835/24, Ansari Road, Daryaganj, New Delhi-110002
Visit us at **www.newagepublishers.com**

PREFACE

One of basic features of technological advancement lies in upgrading the efficiency of production and providing for qualitative advances at the technical level and in the structure of the national economy. The problem of upgrading the efficiency and improving the quality of products is closely linked to the development and introduction of high-efficiency materials, one such important material being lubricating greases that are employed in all types of machinery and are intended to ensure their reliable and durable service.

In spite of a relatively small amount of production of lubricating greases (the world's annual production is about 1 million tons), they are one of the most important and indispensable types of lubricants. This is due to their valuable service properties; their very low specific consumption, sometimes tens and thousands of times less than that of oils; simplification with their use of the designs of machinery, which upgrades their dependability and extends their service life. Moreover, the use of lubricating greases extends the relubrication intervals and greatly cuts down the machinery maintenance costs.

Extensive research has been conducted and a certain degree of progress in improving the quality of high performance lubricating greases and their production processes has been achieved in recent years. The modern assortment includes a broad range of various lubricating grease types. These include greases based on hydrated and anhydrous calcium, ordinary lithium, sodium, barium, zinc, aluminium and other soaps; a special place is occupied by complex lithium, calcium, barium, and aluminium greases; a significant category is the non-soap organically and inorganically thickened greases, which include silica-gel, bentone, carbon black, polymers, polyurea, and other grease types; synthetic hydrocarbon based greases are also of special significance.

The production of such a broad assortment of lubricating greases involves a wide variety of materials of organic and inorganic origin: petroleum and synthetic oils as dispersion media; various natural and synthetic fatty materials and their derivatives, solid hydrocarbons, modified silica gels and clays, carbon black, polymers and pigments, urea derivatives and other compounds, included in the composition of greases or serving as their dispersed phase; and, finally various additions (additives and fillers) which when introduced under certain conditions improve the service properties of greases. Many of these components are produced directly in the grease factories, their production being a component of the grease production process. The knowledge of the composition, methods of production, and other features of the above-listed components is a prerequisite for organization of production of lubricating greases.

A vast range of lubricating greases, a wide assortment of the raw material stock, high demands placed on the quality of the final product and its complex colloidal structure define the features of the grease production process, involving the production equipment used and the development of the automatic process control system as well as the product quality inspection in the process flow. A successful solution of the problems requires a great deal of research and development efforts.

Considerable research on revealing the dependence of properties of lubricating greases on various factors involved in their production process has been conducted over the last two decades. The role and significance of individual stages in the production process in the formation of the colloidal structure and on the properties of lubricating greases have been elucidated. Based on the conducted research, effective engineering solutions as to the production equipment have been found. Problems of a continuous automatic control of the lubricating grease production process and of in-process quality inspection have been solved. All this resulted in development of high efficient production processes providing for production of any type of high-quality lubricating greases.

The author has summarized separate data published by domestic and foreign scientists. The book does not discuss the issues pertaining to the composition, structure, and properties of lubricating greases, as these form the subject of a separate independent publication.

The author will be grateful for the readers' proposals aimed at improving the contents of the book.

Prof. Yu. L. Ishchuk

Foreword

Lubricating grease (or grease, to use a short name), leaving aside synthetic grease for the moment, is arguably amongst the highest "value added" petroleum products. Lubricating greases play very crucial role in the performance of automobiles, industrial machinery and appliances which almost every one of us use, rely upon or work with everyday. Lubricating greases account for about 5% of the global consumption of the lubricants, which in turn account for about 1% of all petroleum products. Thus in volume terms, lubricating greases constitute a relatively small, specialized industry. Nonetheless, the importance of lubricating greases cannot be over-emphasised. The world would quite literally grind to a halt without lubricating greases.

There are not many books on lubricating greases. Even the very few books that are available in English language, largely encapsulate the work and knowledge developed in North America and West Europe. Significant scientific and technological advancements were achieved in the erstwhile "East Bloc" countries during the second half of the 20th century. However, it is only over the last two decades that those outside the erstwhile "East Bloc" have been getting glimpses of the prolific work done by scientists and technologists in that part of the world. This book endeavours to further this process in a small way.

Balmer Lawrie, established in 1867 with Headquarters at Kolkata (formerly Calcutta), commenced manufacture of lubricating greases in 1937, relying entirely on indigenous knowledge. Today as a Company, Balmer Lawrie is the largest manufacturer of lubricating greases in India and possibly one of the global majors in this field. Balmer Lawrie has been interacting with Prof. Ishchuk since the early 90s'. We at Balmer Lawrie have been greatly impressed with the dedicated work that Prof. Ishchuk has carried out particularly in the area of lubricating greases over the last five decades and the phenomenal achievements that he has recorded. Prof. Ishchuk is a titan amongst those who have contributed to technological developments in the field of lubricating greases. We are happy to present this English translation of one of Prof. Ishchuk's several books which we believe would go a long way in meeting a much felt need in Grease manufacturing Technology.

Finally, I wish to acknowledge helpful suggestions and editing support given by my collegues, Shri Abhijit Roy (ARL) and Shri. R. Venkataraman in bringing out the translated work in the present form.

V. Narayan Sharma
Director
Balmer Lawrie & Co.Ltd.

Abbreviations Used

Li-greases	—	lithium greases
cLi-greases	—	complex lithium greases
Na-greases	—	sodium greases
K-greases	—	potassium greases
Ca-greases	—	hydrated calcium greases
aCa-greases	—	anhydrous calcium greases
cCa-greases	—	complex calcium greases
Ba-greases	—	ordinary barium greases
cBa-greases	—	complex barium greases
Zn-greases	—	zinc greases
Al-greases	—	ordinary aluminium greases
cAl-greases	—	complex aluminium greases
Pb-greases	—	lead greasese
LiCa-greases	—	lithium-calcium greases
CaNa-greases	—	calcium-sodium greases
NaCa-greases	—	sodium-calcium greases
Si-greases	—	silica gel greases
Bn-greases	—	bentone greases
HAc	—	acetic acid
HBz	—	benzoic acid
HPt	—	palmitic acid
HSt	—	stearic acid
HOl	—	oleic acid
12-HoSt	—	12-hydroxystearic acid
HRi	—	ricinolec acid
SFA	—	synthetic fatty acids
MSt	—	stearates of metals (Li, Na, Ca, etc.)
MoSt	—	hydroxystearates of metals (Li, Na, Ca, etc)
ATIP	—	aluminium triisopropoxide
DMABAC	—	dimethylalkylbenzylammonium chloride
DMDAC	—	dimethyldialkylammonium chloride
AAODA	—	amino acetic octadecyl
PTFE	—	polytetrafluoroethylene
PTRFE	—	polytrifluoroethylene
PTFCE	—	polytrifluorochloroethylene
PVF	—	polyvinyl fluoride
HOH	—	high-temperature organic heat-transfer agent
GOST	—	USSR state standard (official abbreviation)
OST	—	USSR industry branch standard (official abbreviation)
TU	—	specifications (official abbreviation)
STP	—	works (factory) standard (official abbreviation)

Contents

Page

Preface .... *v*
Foreword .... *vii*
Abbreviations Used .... *ix*

CHAPTER 1. OVERVIEW OF LUBRICATING GREASES 1

1.1. Colloidal Nature of Lubricating Greases 1
1.2. Basic Types of Modern Lubricating Greases and their Classification 6
References 10

CHAPTER 2. DISPERSION MEDIUM OF LUBRICATING GREASES 12

2.1. Petroleum Oils 12
2.2. Synthetic Oils 18
References 23

CHAPTER 3. DISPERSED PHASE OF LUBRICATING GREASES 25

3.1. Soap Thickeners 25
3.2. Solid Hydrocarbons 55
3.3. Inorganic Thickeners 62
3.4. Organic Thickeners 74
References 87

CHAPTER 4. ADDITIVES AND FILLERS 97

4.1. Additives 97
4.2. Fillers 115
References 120

CHAPTER 5. CHARACTERISTIC OF LUBRICATING GREASE PRODUCTION TECHNOLOGY. ITS MAIN STAGES AND THEIR SIGNIFICANCE IN FORMATION OF STRUCTURE AND PROPERTIES OF GREASES 124

References 140

CHAPTER 6. PROCESSING CONDITIONS AND LAYOUTS OF GREASE MANUFACTURING PLANTS, DETERMINATION OF THEIR CAPACITY — 145

6.1. Efficient Layout of Buildings, Arrangement of Equipment, Configuration of Main Process and Auxiliary Flows of Stock Containers, and Finished Products 145
6.2. Plant for Manufacture of Soap Greases by Intermittent Method 148
6.3. Plant for Manufacture of Soap Greases by Intermittent Method with Use of Autoclave-Contactor 150
6.4. Plant for Manufacture of Soap Greases by Semicontinuous Method 152
6.5. Plants for Manufacture of Greases with Finished Air-Dry Soaps by Semi-continuous and Continuous Methods 154
6.6. Plants for Manufacture of Soap Greases by Continuous Method 156
6.7. Plant for Manufacture of Greases Thickened with Highly Dispersed Inorganic and Organic Thickeners 159
6.8. Plant For Manufacture of Hydrocarbon Greases 160
6.9. Determination of Capacity of Lubricating Grease Manufacturing Plants and Shops 161
6.10. Stage-by-Stage Temperature Conditions of Manufacture of Main Types of Soap Greases 164
References 172

CHAPTER 7. SELECTION OF EQUIPMENT FOR INDIVIDUAL STAGES OF LUBRICATING GREASE MANUFACTURING PROCESS — 174

References 197

CHAPTER 8. AUTOMATION OF MANUFACTURING PROCESSES, IN-PROCESS INSPECTION OF QUALITY OF INTERMEDIATE AND OF FINISHED GREASES — 200

References 223

List of Tables

		Pages
1.1.	Some properties of greases of various types	8
1.2.	Grease production trends	9
2.1.	Characteristics of petroleum oils used as dispersion media of greases	13
2.2.	Charaeteristics of synthetic oils as dispersion media of greases	19
3.1.	Characterteristics of metal hydroxides and other compounds, used in manufacture of plastic lubricating greases	26
3.2.	Basic fatty stock used in manufacture of greases	30
3.3.	Saturated fatty acids contained in natural fats	31
3.4.	Unsaturated fatty acids contained in natural fats	32
3.5.	Saturated fatty acids contained in natural fats	34
3.6.	Characteristics of vegetable oils used as saponifiable stock for grease	37
3.7.	Characteristics of ground animal facts	38
3.8.	Characteristics of sperm whale fats	38
3.9.	Characteristics of commercial salomas (OST 18-373-81)	39
3.10.	Characteristics of commercial (stearine) and reactive stearic acids	40
3.11.	Characteristics of commercial oleic acid (GOST 7580-55)	40
3.12.	Characteristics of hydrogenated castor oil and 12-hydroxy stearic acid	41
3.13.	Characteristics of commercial thermally treated and distilled SFA fractions (GOST 23239-78)	44
3.14.	Characteristics of commercial SFA fraction	46
3.15.	Characteristics of commercial petroleum acids (GOST 13302-77)	47
3.16.	Characteristics of colophony of various grades	48
3.17.	Characteristics of distilled tall oil (TU 81-05-26-75)	49
3.18.	Temperature of phase transitions and melting of some soaps	54
3.19.	Characteristics of solid paraffins used as dispersed phase of greases (GOST 23683-79)	57
3.20.	Characteristics of petrolatums (GOST 38-01117-76)	58
3.21.	Group composition of ceresins	58
3.22.	Characteristics of ceresins	60
3.23.	Characteristics of building petroleum bitumens (GOST 6617-76)	61
3.24.	Characteristics of highly dispersed silicon dioxide aerosil (GOST 14922-77)	63
3.25.	Characteristics of aerosil modofied with dimethyldichlorosilane (TU 6-18-185-79)	65
3.26.	Characteristics of highly dispersive silica (STP 3-1-75)	66
3.27.	Characteristics of highly dispersive silica (STP 3-1-75)	66
3.28.	Characteristics of clay minerals	68
3.29.	Characteristics of modifiers	69
3.30.	Characteristics of organobentonites produced in the erstwhile USSR	73

3.31.	Classification of carbon blacks and their production methods	76
3.32.	Physico-chemical characteristics of carbon blacks produced to GOST 7885-77	78
4.1.	Antioxidation additives	99
4.2.	Antiseziure and antiwear additives	104
4.3.	Anticorrosive additives	108
4.4.	Viscous and adhesive additives	112
4.5.	Characteristics of molybdenum disulfide (TI 48-18-133-25)	116
4.6.	Characteristics of graphite produced according to GOST 8296-73	117
4.7.	Characteristics of natural and synthetic micas	119
4.8.	Characteristics of some halogen derivatives of polythylene	120
5.1.	Characteristics of aluminium complex greases manufactured at different component charging sequences	128
5.2.	Dependence of properties of complex calcium greases on thermal treatment temperature and melt holding time	129
5.3.	Influence of cooling method on properties of lithium greases	133
5.4.	Effect of temperature conditions of complex calcium grease manufacture on its properties	136
5.5.	Effect of cooling rate on properties of aluminium complex greases	137
5.6.	Homogenizers and homogenation conditious, recommended for some grease types	139
6.1.	Industry-average yield of reference grease	164
6.2.	Temperature conditions (°C) of manufacture of hydrated and anhydrous calcium greases	165
6.3.	Temperature conditions (°C) of manufacture of calcium-sodium and sodium greases	166
6.4.	Temperature conditions (°C) of manufacture of ordinary lithium greases	167
6.5.	Temperature conditions (°C) of manufacture of complex lithium greases	168
6.6.	Temperature conditions (°C) of manufacture of complex calcium and complex aluminium greases	169
6.7.	Temperature conditions (°C) of manufacture of complex barium greases	170
6.8.	Temperature conditions (°C) of manufacture of aluminium greases	171
6.9.	Temperature conditions (°C) of manufacture of polyurea greases	171
7.1.	Change of properties of petroleum oil-lithium 12-oxystearate system in heating and thermal treatment	174
7.2.	Characteristics of reactors used in manufacture of greases	178
7.3.	Characteristics of scraper-type heat exchangers used in manufacture of greases	181
7.4.	Characteristics of homogenizers used in manufacture of greases	183
7.5.	Characteristics of deaerators used in manufacture of greases	187
7.6.	Characteristics of filters used in manufacture of greases	188
7.7.	Characteristics of mixers used in manufacture of greases	189
7.8.	Characteristics of pumps and metering units in manufacture of greases	192
7.9.	Characteristics of thermic fluid oil AMT-300	194
7.10.	Characteristics of electric heater	195
7.11.	Characteristics of cylindrical tubular furnaces	196
8.1.	Results of reaction mixture's alkalinity and pH of condensate measurements	210
8.2.	Characteristics of capillaries	215
8.3.	Results of "Litol-24" grease viscosity and penetration measurement	219
8.4.	Results of measurements of viscosity and shear strength of grease Litol-24	220

Overview of Lubricating Greases

Lubricants, which include organic and inorganic substances capable of reducing friction, lowering the wear of friction surfaces, and preventing damage by scoring, are employed to upgrade the durability and dependability, serviceability and service efficiency of mechanisms, machines and equipment.

Four basic groups of lubricants are distinguished: gaseous, liquid, plastic, and solid. Most widely used, among them in terms of the variety of products and application conditions, are plastic lubricants, generally known as lubricating greases. In the former USSR they were called consistent greases, which was wrong since the consistency is a property (state) of any substance, including lubricants, which can be gaseous, liquid, grease-like or solid. The term grease is simply used in most cases and will be used further in this book.

1.1. COLLOIDAL NATURE OF LUBRICATING GREASES

Lubricating greases are intermediate in their properties between liquid lubricants (oils) and solid lubricants. They offer the advantages of the two lubricant groups, but substantially differ in the nature and rheological like characteristics from any petroleum and synthetic oils or from solid lubricants like graphite and molybdenum disulfide. Lubricating greases are able to withstand the action of small loads and behave in this case as solids. Thus, at ordinary temperatures they retain the shape imparted to them and do not flow out from small containers when turned upside down, are not thrown out of open lubrication points by centrifugal forces, do not creep down inclined and even vertical surfaces when applied to them as a moderately thick layer, etc. When, however, the load is increased and a critical value, exceeding the ultimate (yield) strength of the plastic system, has been reached, the grease gets deformed and start flowing like ordinary lubricating oils. An important feature of greases is reversibility of this process, the thixotropic recovery: when the load is removed, the flow of grease stops and it acquires the properties of a solid. Most lubricating greases lose the properties of a solid not only under the action of critical loads, but also when the temperature is increased to such a value that the grease resistance, i.e. the ultimate strength, becomes zero.

Based on the above-mentioned features of lubricating greases and based on the definition by V.V.Sinitsyn [1], the following definition of "lubricating greases", relying on their rheological characteristics, can be formulated: a lubricating grease is a lubricant which under the action of small loads at ordinary temperatures exhibits properties of a solid and, when the load has reached a critical value, starts deforming as a plastic and flowing like a liquid; when the load has been removed, it acquires again the properties of a solid. This definition conveys the advantageous feature of lubricating greases over lubricating oils. Another distinguishing feature of greases with respect to oils is their anomalous internal friction: their viscosity does not obey Newton's law; not only is it a function of temperature, but varies also with the shear rate. As the shear rate gradient increases, the viscosity of grease drops steeply, which is a distinguishing advantage of grease over oils.

The main advantages of greases over oils are their ability to hold on in unsealed lubrication points, serviceability over wider range of temperature and speed, better lubricity, higher corrosion protection properties, serviceability in contact with water and other corrosive media and a higher economical efficiency of application. Their drawbacks include a lower cooling capacity, a higher tendency to oxidation, and some complexity of their use in a centralized grease feeding system.

The specific features of lubricating greases, determining the above-mentioned rheological properties, anomalous viscosity and other characteristics are due to their colloidal nature. Let us consider some features of lubricating greases as colloidal dispersion systems.

In the simplest and naturally, somewhat idealized case the greases can be considered as two component system consisting of a liquid base, the dispersion medium and a solid thickener, the dispersed phase. Used as the dispersion medium, which accounts for 75-95% of the grease composition (mass), are petroleum or synthetic oils and other lubricating liquids. The dispersed phase (5-25%) can consist of salts of high molecular weight carboxylic acids (soaps), solid hydrocarbons, inorganic (highly dispersed silica gel, bentonites, lyophilic graphite, etc.) and organic (carbon black, pigments, polymers, urea derivatives, etc.) materials, as well as other highly dispersible substances with a high enough specific surface, capable of getting in nonaqueous media. Such two-component system include only siloxane greases thickened with carbon black, pigments and polymers. In all other cases the greases invariably contain, apart from the dispersion medium and dispersed phase, a third and sometimes even a fourth component, which can be either a mandatory one, without which the lubricating grease does not exist as a stable system, such as water in some soap base greases, or a production-process by product, such as glycerol separated out in saponification of fats, dispersion medium oxidation products formed during heating in the process of manufacture of the grease, etc.

Soap base greases always contain free acids or alkalis, since a stoichiometric saponification of natural fats or neutralization of carboxylic acids is a rare event. Various surfactants are sometimes added to the composition of greases to control the process of formation of structure in their production. Various additives and solid additions, fillers, are employed to improve the service properties of greases. Carbon black, dye base and polymer containing greases become systems containing more than two components. It follows that the greases should be treated as fairly complex polycomponent systems whose basic properties are governed by their composition [1-11]. However, when the greases are evaluated from the view point of the colloid chemistry and their formation processes are studied, they are considered as two-component systems consisting of a dispersion medium and a dispersed phase. As regards the third component, its role in the stabilization of the system under consideration and in imparting its properties is determined in this case [1, 2, 5, 6, 11-22]. The dispersed phase (thickener) forms in the course of grease preparation a three-dimensional skeleton penetrating the dispersion medium throughout its volume. Skeleton elements (dispersed phase particles) have colloidal sizes in two (more often than one) measuring directions. 60-80% of the dispersion medium is held in the cells of the three-dimensional grease skeleton by adsorption bonds, and the remaining part, mechanically. It follows that greases are colloidal dispersions formed by thickeners in a lubricating oil. Their properties as a solid stem from the presence of skeleton, whose properties are determined by the nature, dimensions (dispersity and anisometricity) and shape of dispersed phase particles and by energy bonds between the particles.

The thickener determines the main characteristics of the greases. D.S.Velykovsky [2] classifies thickeners into three basic groups with respect to their nature and interaction with the dispersion medium and thickening mechanism:

- polymorphic thickeners, whose interaction with the dispersion medium strengthens as they change over into high-temperature mesomorphic phases, i.e., thickeners that do not interact with oils at ordinary temperatures and colloidally disperse at elevated temperatures. Such thickeners include soaps.
- thickeners exhibiting no polymorphism but melting at relatively low temperatures and forming homogeneous solutions with the dispersion medium at a temperature exceeding their melting temperature. These include solid hydrocarbons.
- heat-resistant organic and inorganic thickeners not dissolving in the dispersion medium and not undergoing phase transformation with increasing temperature promoting their interaction with the dispersion medium. These include highly dispersible silica gels, carbon black, pigments, etc.

Note that in case of heat-resistant highly disperse organic and inorganic thickeners, when their particles (usually of a spherical shape) come together to the distance of action of molecular forces, aggregate (stick together) into shapeless chains and form a spacial structure of the grease; the formation of a colloidal dispersion occurs mainly through penetration of the dispersion medium into the porous space of an extensive surface of the thickener and its holding there by capillary forces. A smaller contribution to the grease formation is made by adsorption forces that hold the dispersion medium in skeleton cells.

Some features of the formation of structure and the technology of organically thickened greases, data on the formation and properties of soap and hydrocarbon greases depending on the thickener type and grease production technology will be discussed below (Chapters 3 and 5). Since, soaps and solid hydrocarbons are the most widely used thickeners, accounting for over 99% of the total amount of all the thickeners employed, it is appropriate to discuss briefly the specifics of their formation in nonaqueous media and some distinguishing features of these dispersed systems.

Solid hydrocarbons (ozokerite, ceresine, paraffin, petrolatum), melt at temperatures below 100°C and essentially completely dissolve in a petroleum oil at temperatures slightly exceeding their melting temperatures. When such a melt is cooled, crystallization centers (nuclei) appear, crystals grow, a skeleton is formed and the system transforms into a lubricating grease. Dispersed crystalline particles of solid hydrocarbons, formed in the process of cooling, are anisometric. The dispersity and anisometricity of solid phase particles and, naturally, the thickening effect of the hydrocarbon thickener vary depending on the composition of the dispersion medium and of the dispersed phase, presence of the third component and cooling method. Greases produced with these thickeners are hydrophobic, thermally reversible, and exhibit a relatively low temperature limit of plasticity, which is restricted by the melting temperature of the thickener and by its high solubility in the hydrocarbon dispersion medium at elevated temperatures.

The most widespread group of the thickeners are soaps. They practically do not interact with petroleum oils and other lubricating liquids at ordinary temperatures, i.e., when soaps are in a solid crystal phase and undergo no chemical transformation in oils. Under these conditions, soaps neither dissolve nor practically swell in oils; when the temperature is increased and soaps change over into a mesomorphic state (phase transformation), and are able to disperse colloidally in oils. The dispersion of soaps is additionally facilitated and accelerated by a vigorous mechanical stirring or other physical means.

Such a system exhibits different colloidal states depending on the soap concentration in the oil and the temperature. G.V.Vinogradov [16] distinguishes the following basic states of such systems: true solutions or isotropic melts; true gels or simply gels; pseudogels or jellies; suspension or lumps of soap in the solvent. He indicates that the concept of pseudogels was first introduced at the turn of the 20th century by P.P.Weimarn, who noted their characteristic differences from gels. Pseudogels feature dispersed phase particles with an ordered three-dimensional crystal lattice and a skeleton that binds the crystallites and imparts properties of an elastobrittle solid to the system. The gel-to-pseudogel transition is due to the evolution of the crystallization process: two-dimensional liquid crystals, filamentary, platelet, or other linear or branched particles form a three-dimensional skeleton penetrating the pseudogel throughout its volume as a result of growth of dendrites, formation of soap fibers, their joining, mechanical entanglement, and physico-chemical adhesion. The gel-to-pseudogel transformation results in loss of the high elasticity of the system and in strengthening of properties of elastobrittle solids.

Summarizing the above leads to a conclusion that the difference of pseudogels from gels consists solely in the difference in dimensions, namely the length, of dispersed crystalline particles of the thickener. It follows that there exists no fundamental difference between the colloidal state of gels and of pseudogels, and such a division in the light of modern colloid chemistry is undoubtely unreasonable. Therefore the terms gels or oleogels, gel formation rather than pseudogel formation will be used hereinafter.

It should be pointed out that the states of soap-oil systems, depending on their temperature and on the soap concentration, distinguished by G.V.Vinogradov, does not occur in all cases of soap grease formation. They take place only in preparation of greases whose dispersed phase is an alkali metal soap. A lubricating grease is in this case produced by cooling the system from the temperature of an isotropic soap melt in oil down to the temperature of formation of a stable jelly, which then, as a result of a mechanical treatment (pumping, filtering, homogenization, etc.), is broken down and transforms into a plastic oleogel. The system passes through all intermediate states only when the grease is obtained from an isotropic melt. Cases are known where a grease is produced without bringing the system to the isotropic melt state, such as hydrated and anhydrous calcium greases. It is therefore reasonable to divide soap base greases, depending on the soap cation and on the grease production method, into four basic groups: alkali metal (sodium and lithium) soap greases; alkaline-earth metal (calcium and barium) soap greases; aluminium soap greases; and heavy metal (lead, zinc, etc.) soap greases. Every basic group should be subdivided into subgroups of ordinary and so-called complex (this is particularly important for alkaline-earth and aluminium soap greases). Every group of soap grease has its optimum thickener concentration, temperature and other conditions (such as the presence of a structure stabilizer) for the formation of a colloidal system, oleogel. The grease formation processes have been studied in greatest detail and depth for greases based on ordinary alkali metal soaps and on hydrated and anhydrous calcium soaps.

The process of formation of oleogels (greases) is a multistage one and depends on conditions of formation of crystal nuclei (associates and micelles), their growth (aggregation of micelles and formation of supermicellar structures i.e. fibres), and, finally, emergence of a three-dimensional skeleton that endows the grease with plasticity and other properties. Many of these stages proceed concurrently, which prevents studying any one of them individually.

The formation of disperse particles of the thickener and emergence of the spacial structure are considerably affected by the soap type and concentration, composition and properties of the dispersion medium, presence of surfactants, their concentration and nature, soap crystallization temperature, melt cooling rate, stirring rate and other process conditions. The growth of crystals, their aggregation, adhesion and formation of fibers proceed through ultra-thin (monomolecular) solvate layers of surfactants, if present, or interlayers of the dispersion medium, strongly bonded to the surface of disperse particles and being an element of the dispersed phase, i.e., of the skeleton. At the same time a very insignificant fraction of the dispersed phase is dissolved in the dispersion medium and is an integral part of it. The soap thickener can exist in greases in various modifications, various dispersity and degree of anisometricity depending on the above-mentioned factors. It was ascertained by electron microscope studies, X-ray structure analysis, and other methods that elements of the skeleton of soap base greases are anisometric crystalline particles of petal, acicular, platelet, ordinary filamentary, twisted filamentary, bundle-like and other shapes. Their length is of a macroscopic size; width considerably smaller and thickness or diameter of a colloidal size.

Basic rheological properties of the greases stem from their structural features and are determined, on the one hand, by the dispersity and anisometricity of fibres forming the skeleton and, on the other hand, by the bond energy between structural elements of the skeleton and by interaction of the dispersed phase with the dispersion medium. However, the forces acting between soap molecules within a micelle, between micelles within a single structural element (fibre), and in the skeleton as a whole have not been found out and evaluated so far, nor has the energy of interaction of the dispersion medium with the dispersed phase been estimated. This impedes the understanding of the mechanism of formation of the colloidal structure of the grease, its behaviour in storage and operation. The formation of the final structure of greases often occurs after completion of their production process. This has given rise to the concept of 'maturation' of greases, evidenced by a change in their properties with time, consisting in a sudden hardening or thinning of greases under the action of external factors, a spontaneous fluid separation (syneresis), etc. All this indicates an inadequate stability of the colloidal structure of greases.

To sum up the above, plastic lubricating greases in the modern concept of the colloid chemistry can be treated as highly structurized thixotropic dispersions formed by a three-dimensional skeleton of the dispersed phase whose particles (elements) are of a colloidal size in one or two dimensions and of a macroscopic size in the third dimension, while the dispersion medium, lubricating oil, is held in the skeleton cells by capillary, adsorption and other physical bonds.

1.2. BASIC TYPES OF MODERN LUBRICATING GREASES AND THEIR CLASSIFICATION

As stated above, the greases are lubricating liquids thickened by various thickeners and containing small amounts of other functional additives. The thickener concentration in greases is relatively low, but it is the thickener that determines the main service characteristics of greases [1-11, 14, 21]. Because of this, the classification of greases into the following basic groups by the thickener type is widely used:

1. Greases produced with the use of salts of higher carboxylic (fatty) acids i.e. soaps, as the thickener. They are called soap or soap based greases. This group includes the majority of greases both in the variety and in the quantities produced. Depending on the soap cation, soap greases are subdivided into subgroups: based on alkaline metal soaps—lithium (Li), sodium (Na), and potassium (K) greases, of which lithium greases are used most widely and potassium ones are very seldom used; based on alkaline-earth metal soaps-calcium (Ca) and barium (Ba) greases; and based on other soaps—aluminium (Al), zinc (Zn), and lead (Pb) greases. Depending on the composition of the soap anion, most soap greases of one and the same cation are subdivided into ordinary and complex greases. Complex greases primarily include those where the thickeners consist of complex-anion soaps that are an adsorption, or molecular, complex of a soap of a higher fatty acid and a salt of a low molecular weight organic or mineral acid. cLi, cNa, cCa, cBa and cAl-greases are known [1-6].

It should be noted that for Ca-greases, apart from ordinary hydrated ones (cup greases), where the structure stabilizer is water, and complex ones (adsorption complex of a Ca-soap of higher fatty acids, such as stearic, 12-hydroxystearic, SFA (C10-C20), and a salt of a low-molecular organic acid, mainly calcium acetate), are greases conventionally called anhydrous calcium greases (aCa-greases).

Greases based on mixed soaps are sometimes classified into a separate subgroup of soap greases. They contain a thickener that is a mixture of soaps differing by the cation, such as calcium-sodium, lithium-potassium, lithium-calcium, etc. soaps. Greases produced with such mixed soaps are as well called calcium-sodium (CaNa), lithium-calcium (LiCa), lithium-lead (LiPb), etc. greases, that soap cation whose share in the thickener is greater being the first indicated one. However, when carrying out various technical and economic analyses of grease production and application and selecting the grease production method, such greases are included in the group of the predominant soap cation. Soap greases are often, depending on the soap cation used for their production, called synthetic (synthetic fatty acids are the soap anions) or fatty (natural fats or acids separated from them are the soap anions) ones, such as synthetic or fatty cup greases, etc.

2. Greases produced with the use of thermally stable highly disperse inorganic substances with an extensive specific surface as the thickener are called inorganic or inorganically thickened greases. They include greases produced with the use of a highly dispersed modified silicon dioxide: silica gel (Si) greases; organosubstituted clay minerals: bentone (Bn) greases; oleophilic graphite: graphite (G) greases; and other inorganic thickeners, such as asbestos, oleophilic molybdenum disulfide, etc.

3. Greases produced with the use of solid, thermally and hydrolytically stable, highly dispersed organic substances with an extensive specific surface as the thickener; are called organic or organically thickened greases. They are in turn, depending on the type of the organic thickener used, subdivided into carbon black, dye base, polyurea, polymer, and other greases.

4. Greases produced with the use of high-melting (solid) hydrocarbons-paraffin, ceresine, petrolatum, ozokerite, various natural and synthetic waxes-as the thickener, are called hydrocarbon greases. In some countries, such as in the USA, hydrocarbon greases are not included into lubricating greases and therefore are not taken into account in the overall records of the grease production.

Although the thickener does determine the basic characteristics of greases, they can be varied within certain limits for one and the same thickener by selecting the dispersion medium, adding various additives, fillers, etc. to the grease composition.

Greases are often additionally subdivided, according to the composition of their dispersion medium, into petroleum and synthetic oil greases, which can be further subdivided into subgroups, e.g., petroleum greases, into greases based on paraffin-naphthenic oils, aromatic oils, distillate, residual oils, etc., and synthetic oil greases, into greases based on polysiloxanes, esters, polyglycols, synthetic hydrocarbon oils, and on other synthetic fluids.

The governing role of the thickener in determining the grease properties is confirmed by the summarizad data presented in Table 1.1. Besides, the classification of greases by the thickener type is widely employed in technical and economic studies to evaluate the general level of the quality of greases being produced, the structure and amounts of their production, and the development prospects. Table 1.2 presents the structure and history of grease production in some leading industrialised countries and in the former USSR [24-27].

The classification of greases by the type of thickener is not only important to evaluate the grease quality level and to provide a preliminary information on their use, but is also of great importance in determining the grease production technology and in selecting the grease production method and equipment.

Apart from the classification of greases by the thickener type, i.e., by their composition, their classification based on the purpose and functional characteristics is widely used. A detailed information on this and other types of grease classification can be obtained from special literature [28-30]. There is also GOST 23258-78, which specifies the classification of greases by their functional purpose. We will only note here that greases are conventionally divided based on their end use into antifriction greases, which reduce the friction and wear in mechanisms; preservative (protective) greases, which protect metallic products from corrosion; sealing greases, which seal clearances in mechanisms and equipment; and rope greases, used to coat steel ropes.

Table 1.1. Some properties of greases of various types

Grease type	Petroleum-oil greases				Polyethylsiloxane greases		
	Drop point °C[1]	Maximum application temperature, °C	Hydrolytic (water) stability	Antiwear and antiseizure properties	Maximum application temperature, °C	Hydrolytic (water) stability	Antiwear and antiseizure properties
Soap greases							
Sodium	130-160	100-110	Poor	Satisfactory	110-115	Poor	Low
Lithium	175-205	110-125	Good	Satisfactory	120-130	Good	Low
Lithium complex	250	150-160	Good	High	160-170	Good	Satisfactory
Hydrated calcium	70-85	60-70	High	Good	—	—	—
Anhydrous calcium	130-140	100-110	High	Good	—	—	—
Calcium complex	230	140-150	Satisfactory (absorb water and compact)	High	160-170	Satisfactory (absorb water and compact)	Good
Aluminium	95-120	65-70	High	Good	—	—	—
Aluminium complex	250	150-160	High	High	160-170	High	Good
Inorganic greases							
Silica gel	Absent	130-170	Good	Moderate	160-170	Good	Poor
Bentonite	Absent	120-150	Good	Satisfactory	130-150	Good	Poor
Organic greases							
Carbon black	Absent	160-200	High	High	300-350[2]	High	Good
Polymeric (fluorinated hydrocarbons)	Absent	80-150	Satisfactory	High	140-160	Satisfactory	Good
Pigment	Absent	160-200	Good	High	250-300[2]	Good	Good
Polyurea	250	150-200	Good	Good	200-230[2]	Good	Satisfactory
Hydrocarbon greases							
	50-70	50-60	High	Good	50-65	High	Satisfactory

1 The drop point depends insignificantly on the composition of the dispersion medium.
2 Based on polymethylphenylsiloxanes.

Table 1.2. Grease production trends

Grease type	USSR				USA			Canada
	1970 %	1975 %	1980 %	1983 %	1971 %	1975 %	1981 %	1981 %
Sodium	12.20	12.8	6.7	6.5	8.8	5.8	4.2	5.5
Lithium	2.30	6.1	14.6	16.5			56.7	36.8
Lithium complex	—	—	—	Experimental production	52.2	56.2	2.7	23.8
Hydrated calcium	84.70	79.7	77.7	74.9		10.9	8.9	10.5
Anhydrous calcium	—	—	—	Experimental production	23.9	3.8	3.6	0.1
Calcium complex	0.20	0.7	0.4	1.5		6.0	3.5	—
Aluminium	0.20	0.2	0.2	0.2		4.5	1.0	—
Aluminium complex	—	—	—	Experimental production	2.5	3.5	7.3	0.8
Other soap greases	0.39	0.3	0.3	0.3	2.4	1.8	2.5	1.0
Inorganic and organic greases	0.01	0.1	0.1	0.1	10.2	11.0	9.61	21.5
Total	100.00	100.0	100.0	100.0	100.0	100.0	100.0	100.0
Capacity utilised*	100.0	98.0	97.2	85.3	100.0	92.8	115.6	—

*__Note:__ USSR base year 1970; USA base year 1971.

In addition to its main functions, greases also perform other roles : most antifriction greases have protective properties, while preservative and sealing greases can perform functions of antifriction ones, etc. Working/preservative greases, which serve both as preservative and antifriction greases, have found wide use in recent years.

Antifriction greases are divided into general-purpose greases for ordinary temperatures, multipurpose (multifunctional) for elevated temperatures, high-temperature, frost-resistant, automotive, railroad, industrial, special, instrument greases, etc. Sealing greases are divided into valve, thread, vacuum greases, etc.

With respect to the consistency and viscosity level, plastic and semifluid greases are distinguished. Widely used abroad is the NLGI system of grease classification by the consistency class (group) determined from the penetration value:

| NLGI class | 000 | 00 | 0 | 1 |
| Penetration at 25°C | 445-475 | 400-430 | 355-385 | 310-340 |

NLGI class	2	3	4	5	6
Penetration at 25°C	265-295	220-250	175-205	130-160	85-115

If the permissible range of penetration of grease by its specification is intermediate between classes or covers two classes, then a double number, such as 1/2, is used.

REFERENCES

1. Sinitsyn V.V.; The Choice and Application of Plastic Lubricating Greases, Khimiya, Moscow, 1974.
2. Velikovsky D.S.; Poddubny V.N., Vainshtok V.V. and Gotovkin B.D., Consistent Greases, Khimiya, Moskow, 1966.
3. Vainshtok V.V., Fuks I.G., Shekhter Yu.N. and Ishchuk Yu.L.; Composition and Properties of Plastic Lubricating Greases (Review), TsNIITEneftekhim, Moscow, 1970.
4. Ishchuk Yu.L., Kuzmichev S.P., Krasnokutskaya M.E. et al.; The Present and Future Prospects for Advances in Production and Application of Anhydrous and Complex Calcium Greases (Review), TsNIITEneftekhim, Moscow, 1980.
5. Fuks I.G.; Production and Properties of Plastic Lubricating Greases, in: Oil and Gas Processing Technology, Ed. By *Gureev A.A.* and *Bondarenko B.I.*, Khimiya, Moscow, Part 3, pp. 355-382.
6. Ishchuk Yu.L.; Plastic Lubricating Greases, in: Friction, Wear, and Lubrication (Reference Book), Ed. By Kragelsky I.V. and Alisin V.V., Mashinostroenie, Moscow, 1978, Vol. 1, pp. 270-283.
7. Danilov A.M. and Sergeeva A.V.; Ureate Plastic Lubricating Greases (Review), TsNIITEneftekhim, Moscow, 1982.
8. Shulzhenko I.V., Kobzova R.I., Nikonorov E.M. and Bakaleynikov M.V.; Carbon Black Greases (Review), TsNIITEneftekhim, Moscow, 1983.
9. Fuks I.G.; Additions to Plastic Lubricating Greases, Khimiya, Moscow, 1982.
10. Shibryaev B.S. and Fuks I.G.; Technological Surfactants in Soap Base Greases (Review), TsNIITEneftekhim, Moscow, 1983.
11. Ishchuk Yu.L.; Influence of Dispersed Phase Composition on Structure and Properties of Oleogels, Plastic Lubricating Greases, in: Physico-Chemical Mechanics and Lyophilicity of Disperse Systems, Naukova Dumka, Kiev, 1983, Issue 15, pp. 63-75.
12. Varentsov V.P. and Belova M.G.; Swelling and Dispersion in Petroleum-Origin Mineral Oils, Techn. Bull., TsIATIM, 1941, N 8/9, pp. 4-15.
13. Varentsov V.P. and Fedotova A.F.; Solubility of Soaps in Mineral Oils, Ibid., pp. 16-19.
14. Velikovsky D.S.; Production of Consistent Greases, Gostoptekhizdat, Moscow, 1949.
15. Rebinder P.A.; Surface Phenomena in Disperse Systems. Physico-Chemical Mechanics, Selected Works, Nauka, Moscow, 1979.
16. Vinogradov G.V.; Soaps, Solutions, and Soap Gels, Uspekhi Khimii, 1951, Vol. 20, N 5, pp. 533-559.
17. Trapeznikov A.A.; Rheology and Structurization of Oleocolloids, in: Advances in Colloid Chemistry, Nauka, Moscow, 1973, pp. 201-211.
18. Fuks I.G.; Colloid-Chemical Properties and Application of Surfactants, in: Physico-Chemical Fundamentals of Surfactants, Fan, Tashkent, 1977, pp. 5-17.

19. Shchegolev G.G. ; Microstructure and Properties of Aluminium Consistent Greases, Cand. Sci. Thesis, Moscow, 1968.
20. Neuman E. ; Vamos E. And Vermes E., New Views on the Colloidal Structure of Lubricating Greases, Per. Pol., 1972, Vol. 16, N 3, pp. 267-276.
21. Ishchuk Yu.L. and Fuks I.G. ; Modern State and Prospects of Research in Field of Plastic Greases, in: Plastic Greases, Proc. Of 2nd All-USSR Sci. And Techn. Conf. (Berfyansk, 1975), Naukova Dumka, Kiev, 1975, pp. 7-11.
22. Morei Tadashi ; Recent Advances in Consistent Greases, Dzunkatsu, 1980, Vol. 25, N 8, pp. 519-524.
23. Podlennykh L.V. ; Study of Lithium Oxystearate Structurization in Grease Production Process, Cand. Chem. Sci. Thesis, Moscow, 1982.
24. Ishchuk Yu.L. ; State and Prospects of Production and Application of Plastic Lubricating Greases, Neftepererabotka i Neftekhimiya, 1977, N 10, pp. 35-36.
25. Murry I.L. ; NLGI's 1973 Production Survey, NLGI Spokesman, 1975, Vol. 38, N 11, pp. 295-299.
26. Zaborskaya Z.M. and Stolina G.L. ; On Some Trends in Advance of Plastic Lubricating Grease Production in the USA, Neftepererabotka i Neftekhimiya, 1976, N14, pp. 59-63.
27. Quigg J.S. ; 1981 NLGI Grease Survey, NLGI Spokesman, 1983, Vol. 47, pp. 128-134.
28. Sinitsyn V.V. ; Plastic Lubricating Greases in the USSR. Assortment (Reference Book), Khimiya, Moscow, 1979.
29. Sinitsyn V.V. ; Plastic Lubricating Greases Abroad (Reference Book), Khimiya, Moscow, 1983.
30. Papok K.K. and Ragozin N.A. ; Dictionary of Fuels, Oils, Greases, Additives, and Special Liquids, Khimiya, Moscow, 1975.

CHAPTER 2

Dispersion Medium of Lubricating Greases

Various lubricating oils and liquids are used as the dispersion medium of greases [1-5], which accounts for 75-95% of their volume (mass). The overwhelming majority (over 99%) of greases are produced with petroleum oils. Greases produced with polysiloxanes, esters, fluoro- and fluorochlorocarbons as well as on other synthetic oils and liquids are, however, used in some cases to meet specific conditions of operation of individual machines and mechanisms. Vegetable oils, such as castor oil, are in some instances used as the dispersion medium.

Although the most important service characteristics of greases are determined by the thickener type, many of their properties depend nevertheless on the dispersion medium. The nature, chemical, group, and the boiling range of the dispersion medium substantially affect the formation of structure and the thickening effect of the disperse phase and consequently the basic rheological characteristics of greases. Furthermore, the nature and composition of the dispersion medium determine the serviceability of greases in a certain range of temperatures, force and speed, loads, their oxidation characteristics, protective properties, resistance to corrosive media, etc. A considerable role in determining the grease properties is also played by other dispersion medium characteristics. Thus, the volatility of greases depends on the molecular weight and the flash point of their base oils; low-temperature properties of greases (their low-temperature viscosity, resistance to rotation of bearings) depend on the low-temperature viscosity of the dispersion medium, etc.

Many characteristics of the dispersion media depend not only on the nature of the raw stock from which they have been produced, but also on their production method; this is especially valid for petroleum oils. As is well known, commercial petroleum oils of one and the same viscosity level, but produced by different refining methods (e.g., by selective, acid-alkali, and adsorption refining), differ substantially in the group and chemical compositions, which, when they are used as the dispersion medium, significantly affects the grease properties. The use of one or the other dispersion media is also governed by the grease production technology and by the need to meet corresponding fire safety and industrial sanitation regulations.

2.1. PETROLEUM OILS

Industrial or other general-purpose oils are mainly used in the erstwhile USSR for the grease production. Oils specially intended for use as dispersion media of lubricating greases

DISPERSION MEDIUM OF LUBRICATING GREASES

Table 2.1. Characteristics of petroleum oils used as dispersion media of greases

Oil, grade, standard, specifications	Viscosity, mm²/s 50°C	Viscosity, mm²/s 100°C	Flash point in open crucible, °C (not below)	Congelation point, °C (not over)	Refining
1	2	3	4	5	6
I-5A, GOST 20799-75	4.5-5.1	—	120	−25	Acid-alkali
Instrument oil MVP, GOST 1805-76	6.3-8.0	2-2.5	135	−60	Sulfuric-acid
Transformer oil, GOST 982-80 or GOST 10121-76	9.6	2.5-3.5	150	−45	Selective
I-12A, GOST 20799-75	10-14	3-4	165	−30	Selective
Spindle oil AU, GOST 1642-75 or TU 38 101586-75	12-14	3.5-4.0	165	−45	Sulfuric-acid or Selective
Distillate oil fir grease manufacture, grade 20, OST 38 01.120-76	18-22	4-6	170	−15	Alkali
I-20A, GOST 20799-75	17-23	4-6	180	−15	Acid-alkali or selective
ASV-5, TU 38 101158-74	18-23	5±0.3	190	−45	Selective
Spindle oil for cup grease, TU 38 USSR 201-202-75	20-33	5-7	170	−15	Direct distillation of Anastasiev-field petroleum
I-4A, GOST 20799-75	35-45	—	200	−15	Acid-alkali or selective
Distillate oil fir grease manufacture, grade 45, OST 38 01.120-76	40-45	—	180	−8	Alkali
I-50A, GOST 20799-75	47-55	8-9	200	−20	Acid-alkali or selective
Oil-component for Li-greases, Omsk PRW, STP 148-75	65-72	—	200	−15	Selective
VN-70, TU 38 101308-78	71-75	—	230	−10	Selective
Medical vaseline oil, GOST 3164-78	28-36	—	195 (in closed crucible)	−5	Acid-alkali with clay treatment
Cylinder oil 11, OST 38 0185-75	—	9-13	215	+5	Contact-alkalli
Oil for greases, TU 38 USSR 201-227-76	—	9-13	185	−10	Contact-alkalli
High-refined oil for greases, TU 38 20124-76	—	10-13	220	−12	Contact-alkalli
M-11 (DS-11) TU 38 101523-80	—	11±0.5	200	−15	Selective
MS-14, GOST 21743-76	—	≥14	220	−30	Selective

Table 2.1. (continued)

1	2	3	4	5	6
Diesel oil for complex calcium greases, TU 38 USSR 201-296-78	—	≥17	230	−10	Selective
M-20A, TU 38101317-72	—	≥20	225 (in closed crucible)	−15	Selective
MC-20 (MC-20c), GOST 21743-76	150–170	≥20	270	−18	Selective
MK-22, GOST 21743-76	180–200	≥22	250	−14	Acid
PO-28, GOST 12672-77	—	26-30	240	−10	Selective
Transmission oil (Nigrol) TU 38 101539-75:					
winter	—	—	170	−20	Direct distillation residue
summer	—	27-34	180	−5	
Petroleum plasticizer PN-6 (PN-6Sh), OST 3801132-77					Selective
P-40, TU 38 101312-78	—	40±4	270	−10	Selective
Cylinder oil 52, GOST 6411-76 or TU 38 10151-76	—	44-64	310	−5	Sulphuric-acid and selective

are commercially produced only on rare occasions. Petroleum oils are practically employed to produce all types of greases classified both by the thickener type and by their functional purpose. The basic characteristics of petroleum oils used for the grease production in the former USSR are presented in Table 2.1.

A major amount of greases is produced with distillate oils or on their compositions with residual oils, whose viscosity at 50°C varies from 4.5 - 5.1 to 65-120 mm^2/s, both selective- and acid-alkali-refined. A relatively small share of greases, mainly those used in heavily loaded friction assemblies as well as at relatively elevated temperatures, is produced with residual oils, as well as oils refined by various methods, whose viscosity is of the order of 17-64 mm^2/s at 100°C.

A detailed information on the starting raw stock, various methods of production of distillate and residual oils and their physico-chemical properties is presented in [6-10]. We will briefly discuss only some principles of the oil production technology. The general scheme of production of distillate and residual oils is shown in Fig. 2.1. It includes also the preparation of solid hydrocarbons (paraffin, ceresine), associated with the production of oils, which are often used as the disperse phase of greases.

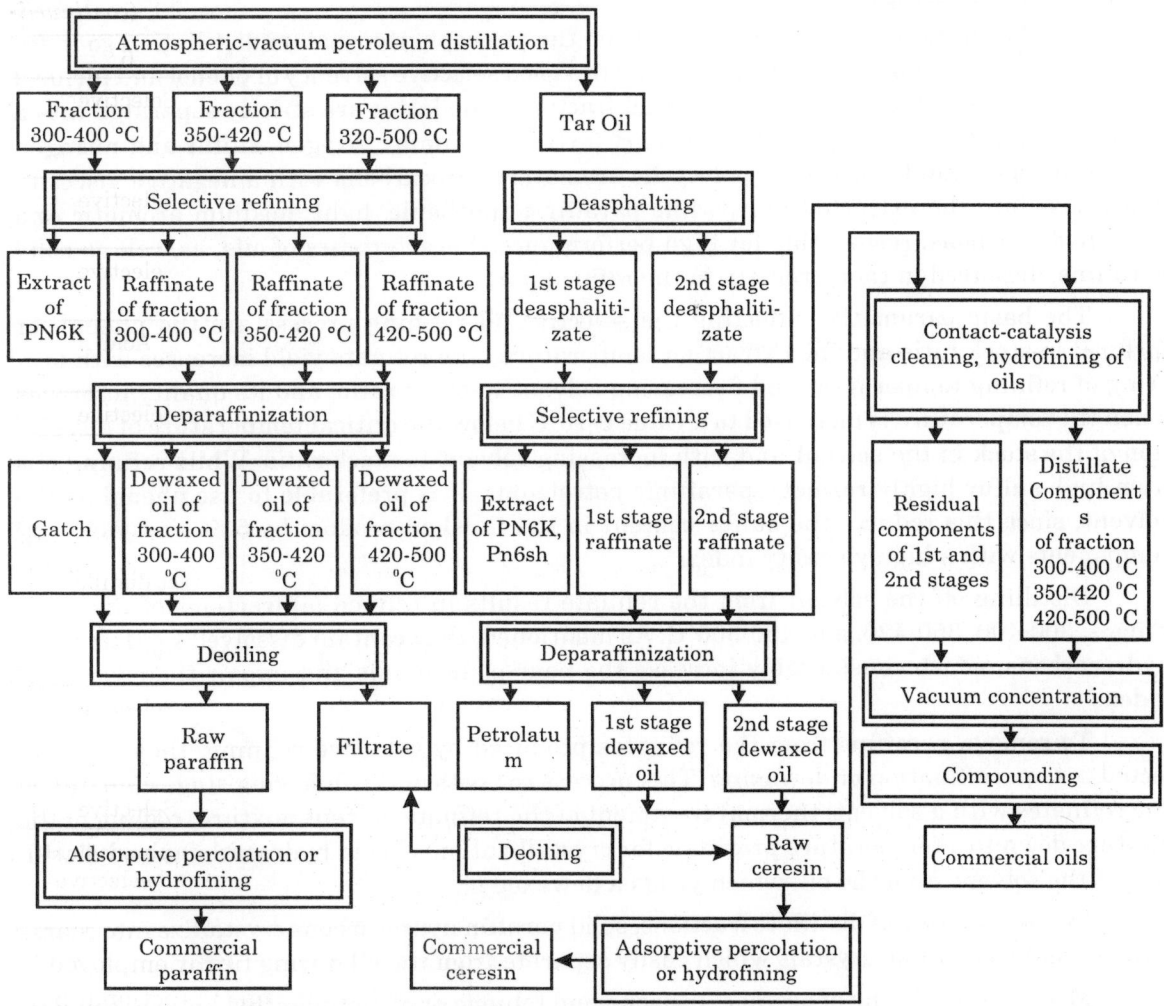

Fig. 2.1. General scheme of production of distillate and residual oil.

Every year the percentage share of oils whose production is based on the acid-alkali refining declines, while the production of selective-refined oils grows. This stems from the fact that modern selective refining techniques make it possible to produce oils from practically any petroleum crude, including those from the eastern fileds, which have a high content of paraffins gummy and asphaltic substances and sulfur. Production of commercial base oils meeting modern requirements (e.g., viscosity index) from such petroleum crudes by the acid-alkali refining, followed by clay treatment, is practically impossible. Moreover, the substitution of the acid-alkali refining by selective oil refining methods eliminates both the problem of utilization of acid sludge and the need to protect the equipment from acid-alkali corrosion.

Oils are produced from petroleum fractions using a two-stream scheme of refining of distillate fractions and residual product, called tar, which is formed by the vacuum distillation of residual fuel oil. The first stream includes refining of three distillate fractions, distilled off within 300-400, 350-420, and 420-500°C (distillate oils of more narrow fractions are used in some cases), and the second, production and refining of residual oils from tar.

The distillate oil fractions resulting from the atmospheric-vacuum distillation go in for selective refining by phenol or furfural. Owing to a good selective solvency of phenol and furfural and their higher density than that of the oil fractions, the latter are strictly separated into a refined part (refinate) and an extracted part (extract). Gummy, oxygen, sulfur and nitrogen containing compounds as well as polycyclic aromatic hydrocarbons with a negative viscosity index pass into the extracted part, while paraffin-naphthenic, light, medium aromatic and other hydrocarbons, responsible for high performance characteristics of oils, as well as solid paraffins, dissolved in them, remain in the refined part.

The basic parameters affecting the selective oil refining process are the phenol- or furfural-to-stock ratio and the extraction temperature. The refinate yield increases with lowering of refining temperature and decreasing solvent-to-stock ratio, and its quality improves when the temperature is increased to a value 8-10°C below the critical temperature of dissolution of the stock in the solvent and with increasing solvent-to-stock ratio. While refining oils from high-sulfur highly resinous paraffinic petroleums, it is preferable to use phenol as the solvent, since this reduces the sulfur content in the finished product by 50% and yields oil components with a high viscosity index.

Distilling off the solvent from the refinate results in refined oil fractions of the same ranges: 300-400, 350-420, and 420-500°C. As mentioned, they contain dissolved solid paraffin hydrocarbons, which considerably increase the pour point of oils and impair their viscosity index.

To remove paraffins from the refinates produced by selective refining, they are subjected to deparaffination or dewaxing. This process consists of the following stages: mixing of the refinates with a solvent; thermal treatment of the refinate-solvent mixture; cooling of the mixture down to a temperature providing for crystallization of solid hydrocarbons and distilling off the solvent from the oil, which yields a dewaxed oil.

In the dewaxing of distillate fractions, solid paraffin hydrocarbons crystallize into coarse platelet and ribbon-like crystals which easily separate from the oil during filtration.

Mixtures consisting of acetone, benzene and toluene or of methyl-ethyl ketone, benzene and toluene are most widely used as solvents in the dewaxing. A benzene-toluene mixture is employed in deep dewaxing.

The main factors affecting the dewaxing process and the final product quality are the ratio between individual components of the solvent, solvent-to-stock ratio, method of stock dilution with the solvent, cooling rate and final temperature of the process. Dewaxed oils resulting from the process are additionally refined by hydrofinishing or by contact refining with clays.

The hydrofinishing process reduces the sulfur and coke contents in oils, improves their colour and susceptibility to additives, upgrades their oxidation and corrosion stability, increases their viscosity index and insignificantly reduces their viscosity. The hydrofinishing of oils is effected with a hydrogen-containing gas at a pressure of about 4 MPa and a temperature of 320-330°C in the presence of a catalyst.

The additional refining of dewaxed oils by clays occurs owing to adsorption properties of appropriately prepared clays, having a high specific surface. The adsorption refining by clays improves the colour and odour of oils, their thermal and oxidation stability, reduces the carbon formation, increases the viscosity index and upgrades other characteristics of oils.

High-viscosity (residual) oils from tars, containing a great amount of asphaltic-resinous compounds, are produced using tar deasphalting by propane as the main process.

Propane at temperatures from -42 to 20°C poorly dissolves paraffins and ceresines. Within 40-60°C, paraffins, ceresines, and the oil part of tar dissolve well in it, while asphaltic-resinous compounds dissolve insignificantly. When the temperature is raised to the critical propane temperature (its boiling point at the deasphalting process pressure is 96.8°C), its solvency for the hydrocarbons useful for oils declines substantially and becomes minimum. These remarkable properties of liquid propane are utilized for the tar separation (fractionation), resulting in two layers: upper, the so-called deasphaltized solution in propane, and lower, an asphalt solution. The deasphalting is carried out in column-type units, tar being fed into the top, and propane, into the bottom part of the column. The solvent-to-stock ratio is from 4:1 to 8:1; the pressure in the unit, 3.8-4.8 MPa; the column botton and top temperatures are recommended to be held at 50-60°C and 75-85°C respectively. Under these process conditions, paraffin-naphthenic and low molecular weight aromatic hydrocarbons pass into the deasphaltized solution, while high molecular weight aromatic hydrocarbons and complex asphaltic-resinous compounds accumulate in the asphalt solution. However, asphaltic-resinous compounds adsorb hydrocarbons of the oil part and carry them over into the asphalt solution. The above process constitutes the first stage of deasphalting, after which the resulting asphalt contains about 30% high index residual oils. To extract these, the second deasphalting stage is used.

In a two-stage deasphalting, the process of the first stage is conducted at a higher temperature, which reduces the deasphalting extraction at this stage, but the overall extraction of oil components at the first and second stages rises considerably (by up to 60%) over that in a single-stage deasphalting and the quality of the oil components produced is better.

The first and second stage extracts from deasphalting serve as starting components for producing motor, aviation, cylinder oils, roll mill oils, and dispersion media for greases intended for severe operating conditions.

The first and second stage extracts from deasphalting are further refined in the same manner as discribed above for distillate oils: selective refining, dewaxing, etc. The dewaxing of first and second stage refinates yields corresponding dewaxed oils and petrolatum which is used as the starting stock to produce petroleum ceresine. It is also used as a component for production of hydrocarbon greases or, after oxidation as flotation reagents, core binders, components of some greases and other useful products.

After the hydrofinishing and additional contact refining by clays, the distillate and residual oil components go in for vacuum concentration and after that compounding (mixing), which yields oils. Commercial oils are also often produced from individual fractions by maintaining appropriate process conditions at the preceding stage. Thus, oils I-5A, MS-6, transformer oil, I-12A are directly produced from the 300-400°C fraction; commercial oils I-20A and ASV-5 from the 350-420°C fraction; oils such as MS-20 and TB-20 from the first-stage and PS-28 and P-40 from the second-stage residual fractions. Practically any commercial oil can be produced by mixing distillate and residual oil components.

Some oils from high-quality oily petroleum crudes (Anastasiev, Suralhany, Baku, etc.) are produced by means of acid-alkali refining and additional clay refining of distillate and residual components. Concentrated sulfuric acid is used in this case to remove unsaturated, sulfur and nitrogen-containing, and resinous-asphaltic compounds. Oxygen and sulfur containing compounds are removed as well as sulfuric acid residues and products of its interaction with hydrocarbons are neutralized and then removed by an aqueous alkali solution.

The acid-alkali refining is less intricate that the selective one. Oils of the same viscosity level, but refined with the aid of selective solvents and with the use of sulfuric acid and alkali, differ markedly in their group composition. The features of the group and chemical composition of such oils, when used as dispersion media of greases, cannot but affect the properties of the latter. Solvent refined oils are in some cases not suitable at all for the production of greases.

2.2. SYNTHETIC OILS

A wide variety of synthetic oils is used as the dispersion medium in production of greases operating over broad temperature and speed ranges, at high unit loads, in a deep vacuum, in a corrosive environment, etc. The following classes of synthetic lubricating fluids are more often used for this purpose: polysiloxanes; esters of saturated fatty and dibasic acids; synthetic hydrocarbon oils; polyalkylene glycols; halogen derivatives of hydrocarbons; polyphenyl ethers, etc. Synthetic oils are employed to produce greases only in cases where petroleum base greases fail to meet the service requirements.

Some data on the synthetic oils used in grease production in the former USSR are presented in Table 2.2. The share of synthetic oil based greases in the total production of all greases in the former USSR is fairly insignificant (within 1%). Most widely used as dispersion media among synthetic oils are polysiloxanes.

Polysiloxanes (polyorganosiloxanes) [11-13] are polymeric compounds consisting of molecules whose main chain contains alternating silicon and oxygen atoms, the remaining silicon valencies being substituted by various organic radicals. In terms of the molecular structure, polysiloxanes are divided into cyclic and linear polymers. Used as dispersion media of lubricating grease are linear polysiloxanes

$$R-\underset{\underset{R}{|}}{\overset{\overset{R}{|}}{Si}}-O-\left[\underset{\underset{R}{|}}{\overset{\overset{R}{|}}{Si}}-O\right]_n-\underset{\underset{R}{|}}{\overset{\overset{R}{|}}{Si}}-R$$

Linear polymeric organosiloxanes are colourless liquids exhibiting a high hydrophobicity, compressibility, chemical and physiological inertness, nontoxicity, stability in corrosive media even at high temperatures, relatively low viscosity variation over a broad temperature range, low volatility and pour point, low surface tension, high dielectric strength and inertness with respect to ordinary rubber. These properties of polysiloxanes depend on the polymeric chain, type and structure of organic radicals (R) attached to the silicon atom. The organic radicals commonly attached are aliphatic and aromatic hydrocarbons. With an aliphatic radical, for example, the thermooxidation stability of polymers degrades with increasing number of carbon

Table 2.2. Characteristics of synthetic oils as dispersion media of greases

Oil, grade	Standart, specifications	Viscosity, mm²/s			Flash point, °C (not below)	Congelation point, °C (not over)	Density, kg/m³
		20°C	50°C	100°C			
Polysiloxane liquids							
PES-S-1	GOST 10957-74	220-230	—	—	265	−70	950-1050
PES-S-2	GOST 10957-74	190-290	—	—	260	−70	950-1050
PES-5	GOST 13004-77	200-500	~100	—	260	−60	990-1020
No.7	GOST 13004-77	42-48	~20	—	190	−70	960-980
PMS-60	GOST 13032-77	57-63	—	—	300	−60	—
PMS-300	GOST 13032-77	285-315	—	—	300	−60	980-1000
PMS-100 r	GOST 13032-77	95-105	—	—	300	−100	—
Copolymer-3		220-400	—	≥ 30	280	−60	—
PMFS-4	GOST 15866-70	450-1000	—	≥ 28	300	−20	1000-1100
Esters							
B-3V	TU 38 101295-75	60-70	17-20	≥ 5	235	−60	990-997
Ether No. 2	TU 38 101272-72	60-70	17-20	4.5-5.3	230	−50	—
DOS thermostable	GOST 19096-73	20-30	7-10	≥ 3	215	−55	≥913
DOS	GOST 8728-77				215		913-919
Hydrocarbon oils							
MAS-14N	GOST 21791-76	-	60-70	8-12	215	−50	—
MAS-35	TU 38 101164-78	-	250-270	≥ 26	270	−32	—
M-9S	TU 38 401269-82	-	250-270	9±0.5	200	−55	—
Other oils and liquids							
Polypropilene glycol			70-90	—	225	−40	—
Polyalkylene glycol		20-200	—	5-40	—	−60	—
Liquid 161-44	TU 6-02-910-74	38-45	—	8.3	250	−90	1020-1110

atoms in the radical. Thus, polysiloxanes containing a methyl radical are thermally stable up to 180-200°C; an ethyl radical, up to 150-160°C; and a butyl radical up to 135-140°C. Polyarylsiloxanes exhibit a higher thermal and oxidation stability than do polyalkylsiloxanes. Thus polyphenylsiloxanes are thermally stable up to 300-350°C. Polyalkylarylsiloxanes occupy an intermediate position. At the same time the presence of an aromatic radical in the compound raises its pour point and substantially impairs the viscosity-temperature characteristic of the polymer. Polyalkylsiloxanes exhibit poor (inadequate) lubricating properties, while polyalkylarylsiloxanes and polyarylsiloxanes, particularly polyhaloorganosiloxanes, offer higher lubricating properties.

Polyethyl-, polymethyl-, and polymethylphenylsiloxanes find the widest use in industrial practice, including in the production of special greases. Characteristics of polysiloxanes used as dispersion media of greases are presented in Table 2.2.

Polyethylsiloxanes (PES) are mainly linear polymers of a general formula

$$C_2H_5-\underset{\underset{C_2H_5}{|}}{\overset{\overset{C_2H_5}{|}}{Si}}-\left[-O-\underset{\underset{C_2H_5}{|}}{\overset{\overset{C_2H_5}{|}}{Si}}-\right]_n-O-\underset{\underset{C_2H_5}{|}}{\overset{\overset{C_2H_5}{|}}{Si}}-C_2H_5$$

where $n = 3 - 12$.

A distinguishing feature of polyethylsiloxanes is their ability to mix in any ratio with petroleum oils. This greatly improves the lubricating properties of polyethylsiloxanes, and a synergistic action occurs at certain polysiloxane/petroleum oil ratios; such mixtures exhibit higher antiseizure and antiwear properties than those of their components. Due to this, polysiloxanes are used as dispersion media not only in a pure state, but also in mixtures with petroleum oils.

Polymethylsiloxanes (PMS) are linear polymers of a general formula

$$CH_3-\underset{\underset{CH_3}{|}}{\overset{\overset{CH_3}{|}}{Si}}-\left[-O-\underset{\underset{CH_3}{|}}{\overset{\overset{CH_3}{|}}{Si}}-\right]_n-O-\underset{\underset{CH_3}{|}}{\overset{\overset{CH_3}{|}}{Si}}-CH_3$$

where $n = 3 - 700$.

· PMS are very seldom used as the dispersion medium of greases. They differ from polyethylsiloxanes and other polymers by a less sloping viscosity-temperature curve and lower lubricating characteristics. They cannot be recommended for lubrication of steel-steel, steel-cast iron and like sliding friction pairs.

Polymethylphenylsiloxanes (PPMS) are linear polymers of a general formula

$$R_aR'_{3-a}Si-\left[-O-\underset{\underset{CH_3}{|}}{\overset{\overset{CH_3}{|}}{Si}}-\right]_m-\left[-O-\underset{\underset{CH_3}{|}}{\overset{\overset{CH_3}{|}}{Si}}-\right]_n-O-SiR_aR'_{3-a}$$

where $R = CH_3$, $R' = C_6H_5$, $a = 0 - 3$, $m = 1 - 15$, $n = 0 - 20$.

PPMS are used to produce special high-temperature greases. Their pour and flash points, thermo-oxidation stability, volatility, solubility and miscibility with other polysiloxanes vary with the number of phenyl radicals in the polymer molecule and their ratio to the number of methyl radicals. The lubricating characteristics of PPMS are superior to those of PMS and PES.

In addition to the above-mentioned polyalkyl- and polyalkylphenylsiloxanes, also polyorganosiloxanes with a halogen in the organic radical can be used as dispersion media for greases. These are oligomers containing chlorine in an aromatic radical or fluorine in an aliphatic radical. Polyhalo-organosiloxanes are colourless liquids insoluble in water, but soluble in aromatic and chlorinated hydrocarbons. Those polyhalo-organosiloxanes which can be used as dispersion media for greases have the following properties: viscosity at 20°C 38 to 120 mm^2/s; flash point 250 to 270°C; pour point -70 to -90°C; and density 1070 to 1110 kg/m^3. They exhibit better antiseizure and antiwear properties than other polysiloxane liquids.

Esters [12-14] which can be employed as dispersion media of lubricating greases are primarily products of esterification of polyhydric alcohols and monobasic saturated carboxylic acids or products resulting from interaction between monohydric high molecular weight alcohols and dibasic carboxylic acids. The former group includes oil B-3V and ester No. 2, which are pentaerythritol esters of a C7-C9 fatty acid mixture and the latter disooctylsebacate (DOS) whose characteristics are given in Table 2.2.

The use of esters as dispersion media makes it possible to produce greases, such as lithium base greases, serviceable over a broad temperature range (from -60 to 130°C). They also offer a good lubricity, but are not serviceable in contact with water, since this will lead to hydrolysis of esters, and the formed mixture has a high corrosion activity. In addition, ester-based greases cause swelling of rubber seals.

Esters are inferior to polysiloxanes in the viscosity-temperature characteristic, thermo-oxidation stability, volatility, resistance to corrosive media and temperature limits of application, especially the upper limit of temperature, but are superior to them in lubricity. They mix well with petroleum oils, which allows the use of such mixtures as dispersion media for greases.

Synthetic hydrocarbon oils [12-18] are prmarily produced by polymerization of olefins or alkylation of aromatic hydrocarbons by olefins, which yields polyolefins and alkylated aromatic hydrocarbons respectively.

The starting stock for producing polyolefins are ethylene, propylene, butylene, forming as the pyrolysis of ethane, cracking of paraffins, hydration of acetylene and other methods.

Oils based on alkylated aromatic hydrocarbons include alkyl, dialkyl, and polyalkyl deriatives of benzene, naphthalene, diphenyl and other aromatic hydrocarbons. They exhibit good viscosity-temperature characteristics, low pour point and other valuable properties. Note that the so-called straight chain alkyl benzenes offer lower pour points, a better viscosity index and a higher thermal stability compared to branched ones. The use of alkyl benzenes as the dispersion medium allows the production of greases, such as bentone and carbon black greases or greases based on other thermally stable thickeners, applicable over a temperature range from -60 to 200°C. These greases also offer good antiwear properties and has little effect on seals of ordinary rubber and plastics [17, 18].

Production of greases based on alkyl benzene-type synthetic hydrocarbon oils is very limited due to high cost and scarcity of these oils. The development and production of synthetic hydrocarbon oils open up prospects for creation of highly effective low-temperature greases and greases serviceable over broad ranges of temperatures, speeds and loads.

Mixtures of synthetic hydrocarbon oils with petroleum oils can also be used to produce greases. It should be noted that the research on employment of synthetic hydrocarbon oils and their mixtures with petroleum oils for production of greases has been limited and scientifically substantiated requirements for the structure of hydrocarbons and properties of these oils as dispersion media are so far absent.

Synthetic dispersion media of special-purpose lubricating greases can also be polyalkylene glycols, polyphenyl ethers, fluoro- and fluorochlorocarbons, perfluoroalkyl amines and other similar lubricating liquids [1, 2, 11-14, 19].

Polyalkylene glycols, known also under the names of polyglycols, polyethers, polyoxyalkylene glycols and polyglycol oils are linear polymers of a general formula

$$RO - \left[-CH_2 - \underset{\underset{R}{|}}{CH} - O - \right]_n - R$$

where R = identical or different organic radicals or hydrogen.

Polyglycol oils are produced by alkylation of polyhydric alcohols. Depending on the structure of the radical R, they are either good or poorly soluble in water or not soluble in it at all. If R-hydrogen, methyl or ethyl radicals, then such polyglycol oils are water-soluble; they are more often used as cutting fluids. But if R = propyl, butyl or a higher molecular weight radical, then such polyalkylene glycols are poorly soluble in water or practically not soluble in it at all. These polyglycol oils are of interest as dispersion media for greases.

Polyglycol oils differ from petroleum oils in terms of lower volatility, gum formation and fire hazard, higher viscosity index, thermal conductivity, and antiwear properties, a lower pour point and are relatively inert to rubber. They are suitable as dispersion media for semifluid greases intended for lubrication of various speed reducer gearings.

Polyglycol oils with various viscosities can be produced. Oils with a viscosity at 100°C from 2.5 to 70 mm^2/s have a viscosity index from 135 to 180 and a pour point from -29 to -70°C. They also exhibit higher flash (200-260°C) and self-ignition (340-400°C) points than do petroleum oils of same viscosity. The ignition point of polyalkylene glycols is 10-40°C higher than their flash point. These properties can be significantly upgraded by addition of antioxidants to polyglycols. The density of polyglycol oils at 25°C is within 950-1220 kg/m^3.

Polyphenyl ethers (polyphenyl oils) are synthesized from halogen derivatives of aromatic hydrocarbons and phenolic compounds; their molecule contains four to six benzene rings, and ether bonds are in the m-position. Some of the valuable properties of polyphenyl oils are their exceedingly high stability under the action of high temperatures, oxygen and radioactive radiations. It is for such extreme conditions that their use as dispersion media of greases is preferred; an appropriate thickener is to be employed in such cases. Using such oils to produce soap base greases is difficult.

Apart from the above unique characteristics inherent in polyphenyl ethers, they offer lower fire hazard, better lubricating properties, higher viscosity index and lower pour point compared to petroleum oils. Their viscosity at 100°C is within 3-69 mm^2/s.

Polyphenyl oils containing alkyl groups start thermally decomposing at 350-380°C, and oils containing no such groups, decompose only at 440-465°C. In presence of such metals as copper, iron, aluminium, magnesium, titanium, etc. they are not oxidized by the oxygen in air at 300-320°C. Their self-ignition temperature is in the range of 550-590°C.

Fluoro- and fluorochlorocarbon-based oils are produced on a commercial scale by halogenation of kerosene and other petroleum fractions.

Properties of fluoro and fluorochlorocarbons depend on the selection of not only the desired petroleum fraction, but also the fluorinating agent: manganese or cobalt trifluoride, etc. In appearance, fluorocarbon oils resemble petroleum oils; they are mainly light-yellow liquids. In other physico-chemical properties, however, fluorocarbon oils substantially differ from petroleum oils. Their density is in the range of 1900-2000 kg/m^3. They are thermally stable up to 400-500°C, neither ignite nor burn, are stable under the action of strong acids, and other corrosive media, do not oxidize, do not corrode various metals, have high lubricating properties, a high volatility, a poor viscosity-temperature characteristic, a high pour point, and a low (from -50 to -400) viscosity index.

They are used to produce special greases, most often thickened with solid polymers, such as polytetrafluoroethylene, polytrifluorochloroethylene, etc. These greases are employed in contact with liquid and gaseous oxygen, hydrogen peroxide, nitric acid, halogens and other corrosive media.

A wide use of fluorocarbon oils in formulations of greases is inhibited not only by their poor viscosity-temperature characteristics and high volatility, but also by their high cost, hundreds of times that of petroleum oils.

Used as dispersion media of special greases for applications in corrosive media, apart from fluorocarbon oils, produced from petroleum fractions, are also perfluorocarbons (perfluoroalkanes), hydrocarbons where hydrogen atoms are substituted with fluorine. Their boiling point, despite a high molecular weight, is below that of their hydrocarbon analogues. They are superior to noble metals in the stability under the action of oxidants and other corrosive media, such as aqua regia. Perfluorocarbons do not oxidize under the action of oxygen right up to their thermal decomposition temperature, do not hydrolyse when repeatedly treated with a 20% aqueous solution of sodium hydroxide at 100°C and very weakly react with a 40% alcoholic solution of potassium hydroxide as well as with molten alkali metals. Perfluorocarbons practically do not corrode metals. Apart from the above-mentioned synthetic oils, other synthetic dispersion media, such as perfluorotrialkyl amines, etc., can be used in the production of greases.

REFERENCES

1. Boner K.D. ; Production and Application of Greases. Transl. from Englysh, Ed. by Sinitsyn V.V., Gostoptekhizdat, Moscow, 1958, 104 p.

2. Sinitsyn V.V. ; The Choice and Application of Plastic Lubricating Greases, Khimiya, Moscow, 1974, 416 p.
3. Velikovsky D.S. ; Poddubny V.N., Vainshtok V.V. and Gotovkin V.D. Consistent Greases, Khimiya, Moscow, 1966, 264 p.
4. Vainshtok V.V., Fuks I.G., Shekhter Yu. N. and Ischchuk Yu.L. ; Composition and Properties of Plastic Lubricating Greases (Review), Ts NIIT Eneftekhim, Moscow, 1970, 80 p.
5. State Standards of the USSR: Greases, Izd. Standartov, Moskow, 1982, Part 1.
6. Chernozhukov N.I. ; Krein S.E. and Losikov B.V. ; Chemistry of Mineral Oils, Gostoptekhizdat, Moskow, 1959, 415 p.
7. Chernozhukov N.I. ; Oil and Gas Processing Technology, Ed. by Gureev A.A. and Bondarenko B.I., Khimiya, Moscow, 1978, Part.3.
8. Goldberg D.O. and Krein S.E. ; Lubricating Oils from Eastern-Field Petroleums, Khimiya, Moscow, 1982, 232 p.
9. Glazov G.I. and Fuks I.G. ; Production of Petroleum Oils, Khimiya, Moscow, 1976, 192 p.
10. Korelyakov L.V. and Shkolnikov V.M. ; Modern High-Index Oils from Petroleum Stock, TsNIITEneftekhim, Moskow, 1972, 52 p.
11. Muzovskaya O.A., Yakusheva T.M. and Pospelova S.G. ; Organosilicon Products Fabricated in the USSR. Catalogue-Reference Book, Khimiya, Moscow, 1970, 52 p.
12. Synthetic Lubricating Oils and Fluids, Ed. by Hunderson R.S. and Hart A.V. Transl. from English, Ed. by Vinogradov G.V., Khimiya, Moscow-Leningrad, 1965, 385 p.
13. Papok K.K. and Ragozin N.A. Z ; Dictionary of Fuels, Oils, Greases, Additives and Special Liquids, Khimiya, Moscow, 1975, 392 p.
14. Dintses A.I. and Druzhinina A.V. ; Synthetic Lubricating Oils, Gostoptekhizdat, Moscow, 1958, 352 p.
15. Tsvetkov O.N., Shkolnikov V.M. ; Bogdanov Sh.K. and Toporushcheva R.I., Poly-a-Olefin-Based Lubricating Oils. Khimiya i Tekhnologiya Topliv i Masel, 1982, N 10, pp. 42-44.
16. US Patent N 488, 622. Preparation of Synthetic Hydrocarbon Lubricants. Guire S.E., Riddle I.L. ; Nicks G.E. et al. Published September 30, 1975.
17. Kinau E.R. and Woods W.W. ; Dialkylbenzene-Based Lubricants. Khimiya i Tekhnologiya Toplivi Masel, 1971, N 5, pp. 35-38.
18. Scott W.P. and Cloud A.P. ; Dialkylbenzene Based Lubricants for Extreme Temperature Service. NLGI Spokesman, 1977, Vol. 40, N 8, pp. 260-264.
19. Zaslavsky Yu.S. ; Radiation Stability of Lubricants, Gostoptekhizdat, Moscow, 1961, 160 p.

CHAPTER 3

Dispersed Phase of Lubricating Greases

The dispersed phase, despite its relatively low (5-25%) concentration in a grease, governs basic service characteristics of greases. Used as the disperse phase of greases most often are salts of higher fatty acids, soaps, and solid hydrocarbons; and less often, inorganic origin thickeners; and in small amounts, organic thickeners [1,2].

3.1. SOAP THICKENERS

Salts of higher fatty acids, employed in grease production, are called soap thickeners. Greases are known whose thickeners are lithium, sodium, calcium, magnesium, potassium, zinc, strontium, barium, aluminium, lead and other soaps. However, an extensive use has been found of lithium, sodium, calcium, barium, and aluminium greases, while potassium, lead and zinc soaps are much less used to produce greases. Mixed soap-based greases, such as calcium-sodium, lithium-calcium, lithium-potassium, calcium-lead, etc. are also often used.

In the USA [1,3,4], 50-60% of soap greases are produced with soaps of individual fatty acids, stearic and 12-hydroxystearic acid. Hydrogenated castor oil, consisting of 80-82% of glycerides of 12-hydroxistearic acid, is used in many cases, especially in production of lithium greases. The remaining soap greases are prepared with soaps of natural fats. Up to 60% of the total consumption of natural fats are fats of ground animals; 25-30%, to fats of sea animals and fish; and 10-15%, vegetable oils. In the former USSR, synthetic fatty acids (about 65%), vegetable oils (about 20%), fats of ground and sea animals as well as hydrogenated natural oils and fats (called «salomases» in the former USSR) (about 5%) and individual commercial natural acids (stearic, 12-hydroxystearic, and oleic) (up to 10%) are used for this purpose. The composition and properties of soaps, depending on their cation and fatty-acid radical (anion), determine to a great extent the performance characteristics of greases. Due to this, it is essential to know the basic properties of both cations and anions of soaps as well as of soaps themselves and their production methods.

3.1.1. Soap Cation

Alkali metals of the first group of the periodic table (Li, Na, K), alkaline-earth metals of the second group (Ca and Ba), a metal of the third group (Al), and a metal of the fourth group (Pb) are widely used as soap cations in the dispersed phase of greases; hydroxides or other compounds of these metals are being employed to produce preformed soaps or as in situ soaps formed during the grease production process. Characteristics of such compounds are presented in Table 3.1 [5, 6].

Table 3.1. Characterteristics of metal hydroxides and other compounds, used in manufacture of plastic lubricating greases

Compounds	Formula	Appearance, crystal structure, and other characteristics	Molecular mass	Density, kg/m³	Melting point, °C	Interaction with water	
						cold	hot
1	2	3	4	5	6	7	8
Lithium hydroxide	$LiOH \cdot H_2O$	White monoclinic crystals	41.96	1830	Over 600 (with H_2O removal)	Dissolves at 10°C: 23 g in 100 g	at 80°C: 26.8 g 100 g
Sodium hydroxide	$NaOH \cdot nH_2O$ (n = 1-5)	White acicular or monoclinic crystals	58.01-174.10	64.3-7.6 (decreases with incre-asing "n")	Well soluble		
Potassium hydroxide	KOH	White deliquescent crystals	56.11	2044	400.0	Well soluble at temperature over 95°C	
Calcium oxide	CaO	Colorless cubic crystals. Externally white fine loose powder	56.08	3370	2580.0	Reacts to yield $Ca(OH)_2$	

DISPERSED PHASE OF LUBRICATING GREASES

Calcium hydroxide	$Ca(OH)_2$	Colorless acicular crystals. Externally white fine loose powder	74.09	2240	580	Poorly dissolves at 20°C: 0.166 g in 100 g; at 100°C: 0.08 g in 100 g
Barium hydroxide	$Ba(OH)_2 \cdot 8H_2O$	White monoclinic crystals. Externally white fine powder. Toxic, fire and explosion hazardous	315.48	2180 (at 16°C)	78	Dissolves at 15°C: 5.6 g in 100 g; in boiling water: very well
Aluminium hydroxide	$Al(OH)_3$	White monoclinic crystals. Exhibits amphoteric properties	78.00	2420	180-200	Not soluble, forms voluminous jelly-like white sediment
Potassium alum	$KAl(SO_4)_2 \times 12H_2O$	Colorless complex salt; octahedral crystals, well dissociated in water. Externally white fine-crystalline powder	474.39		93.4 (H_2O is removed as temparature rises)	Dissolves at 15-30°C: 5-9 g in 100 g; at 100°C: 154 g in 100 g
	$Al_2(SO_4)_3 \times K_2SO_4 \times 24H_2O$	Colorless complex salt; cubic or monoclinic crystals, well dissociated in water. Externally white fine-crystalline powder	948.78	1750	64.5 (with removal of $18H_2O$)	Dissolves at 20°C: 11.4 g in 100 g; at 85°C: 280 g in 100 g
Aluminium isopropylate	$Al(OC_3H_7)_3$	Colorless liquid, distilling at 140-160°C and 666.5 Pa (5 mm Hg). Slowly solidifies into white crystalline substance	204.00	900 (at 60°C)	118.5	Reacts, hydrolyzing to $Al(OH)_3$. Stored in organic solvents: alcohol, gasoline, oils
Lead acetate	$Pb(CH_3COO)_2 \times nH_2O$ ($n = 1; 3; 10$)	White orthorombic or monoclinic crystals. "Saturn sugar", $Pb(CH_3COO)_2 \times 3H_2O$, is more often used in technology	379.33	2550	22.0	Well soluble; Very well soluble

Li-, Na-, and K-greases are produced with the use of aqueous solutions of hydroxides of respective alkali metals, which are strong alkalis. The reaction is conducted by a direct saponification of fats or neutralization of acids.

Lithium greases are prepared with the use of commercial lithium hydroxide LiOH . H_2O, produced according to GOST 8595-75. Under this GOST, three lithium hydroxide brands are produced: LGO-1, LGO-2, and LGO-3, all containing 53-54% LiOH. A distinguishing feature of the brands is that they contain carbonates of various metals as well as chlorides and sulphates.

Commercial sodium hydroxide according to GOST 2263-79 is produced in several brands: R—solid mercuric; TKh—solid chemical; TD—solid diaphragmic; RR—mercuric solution; PKh—chemical solution; and RD—diaphragmic solution. The words «mercuric», «chemical», «diaphragmic» refer to their production methods. Each of the brands is divided into grades, such as RD of the highest or the first grade. The brands differ in the content of sodium hydroxide and various impurities, such as sodium carbonate, silicic acid, etc. The percentage of NaOH by mass in solid brands is within 94-98.5 and in liquid ones, within 44-46%. NaOH of the TD and RD brands is primarily used in grease production.

Commercial potassium hydroxide is produced according to GOST 9285-78 by a diaphragmic electrolysis of potassium chloride solution. According to the GOST, two basic brands are produced: solid (highest, first and second grades) in the form of a fusion cake or flakes and liquid (highest and first grades). They differ in the percentage by mass of caustic alkalis (KOH + NaOH), which for solid brands varies within 93-95 and for liquid ones, within 52-54 and in the content of various impurities, namely potassium and sodium carbonates calcium, aluminium, chlorides, sulfates, and other components, whose percentage is very insignificant.

Grease production often involves the use of strong aqueous solutions of NaOH and sometimes also of KOH, whose concentration is commonly expressed in per cent by mass or characterized by their density. To convert one into the other it is useful to know that the first two digits after the decimal point in the density value approximately correspond to their percentages [6]:

Content, %	5	10	15	20	25	30	35	40	45	50
NaOH density, g/cm^3	1.055	1.11	1.17	1.22	1.28	1.33	1.38	1.43	1.48	1.53
KOH density, g/cm^3	1.045	1.09	1.14	1.19	1.24	1.29	1.34	1.40	1.46	1.51

Ca- and Ba-greases are as well produced with the use of hydroxides of these metals. These are strong alkalis, and accident prevention rules should be observed in handling them.

Calcium hydroxide («slaked lime») is obtained from calcium oxide by its interaction with water. The lime slaking involves the liberation of a high amount of heat, 66.9 kJ/mole (16 kcal/mole), which should be taken into account in producing calcium hydroxide directly in grease factories. The water solubility of $Ca(OH)_2$ increases and that of $Ba(OH)_2$, declines with decreasing temperature. It should be remembered that $Ba(OH)_2$ is toxic, and poses fire hazard while $Ca(OH)_2$ is absolutely non-hazardous.

Commercial calcium hydroxide is produced according to TU 6-18-75-75 in the form of a fine loose white powder of a definite fractional composition, in three grades (first, second, and third). The percentage by mass of CaO in terms of the hydroxide varies within 91.5-96. An insignificant content of impurities, such as calcium carbonate, magnesium salts, substances insoluble in hydrochloric acid is specified. Lime of the first grade should be used in grease production. It is employed in an ordinary form i.e., as a powder, or as milk of lime, prepared from the powder. A lime-oil suspension, prepared from petroleum oils and slaked lime, is often used.

Commercial barium hydroxide $Ba(OH)_2 \cdot 8H_2O$ is produced on a commercial scale according to GOST 10848-79 in two grades (first and second), where the percentage by mass of $Ba(OH)_2 \cdot 8H_2O$ are 97 and 95 respectively. The content of barium acetate in either of the grades should not exceed 1%; the balance are impurities, such as substances insoluble in hydrochloric acid, chlorides, other barium and calcium salts, etc. $Ba(OH)_2 \cdot 8H_2O$ of the first grade is to be used in grease production.

Aluminium greases are most often produced with the use of potassium alums $AlK(SO_4)_2 \cdot 12H_2O$, $Al_2(SO_4)_3 K_2SO_4 \cdot 24H_2O$, which are complex substances, dissociating well in water. They are neither toxic nor pose any fire or explosion hazard. Commercial alums are produced according to GOST 15028-77 in two grades (highest and first) where the percentage by mass of Al_2O_3 should exceed 10.7 and 10.4 respectively. These grades also differ in the percentage of iron, arsenic and water-insoluble residue. The first grade is used for production of greases.

Aluminium isopropoxide (AIP) is obtained by dissolution of metallic aluminium in isopropyl alcohol in presence of catalysts such as mercuric chloride, or anhydrous aluminium chloride. To prevent its reaction with the moisture in air, it is kept in an organic solvent, such as alcohol or gasoline, as well as in oil medium.

Lead soaps for the subsequent production of greases are produced by a double exchange reaction with the use of lead acetate $Pb(CH_3COO)_2 \cdot 3H_2O$, which is obtained through interaction of PbO with acetic acid and should in its properties correspond to GOST 1027-67 («pure» grade). Lead acetate is in the form of colourless crystals of a sweetish taste, well soluble in water and glycerol; at 75°C and above this temperature it transforms into an anhydrous variety, i.e., loses the three water molecules and in air it weathers to form lead carbonate. It is toxic, belongs to the highest class of hazardous substances, its maximum permissible concentration (MPC) is 0.01 mg/m^3.

3.1.2. Soap Anion (Fatty/Acid Radical)

Soaps, the dispersed phase of greases, are produced from a broad spectrum of fatty materials: fats of ground animals, vegetable oils, fats of fish and sea animals, commercial fatty acids, natural waxes, petroleum acids and other commercial fatty materials. In addition, various fractions of synthetic fatty acids (SFA), obtained by oxidation of paraffin, as well as oxidized petrolatum were employed in the former USSR. This fatty stock is called the saponifiable component of greases. The main saponifiable components used in the former USSR are presented in Table 3.2.

The composition and properties of the saponifiable stock determine to a great extent the basic characteristics of soap greases [7], and therefore it is essential to know the properties of fats or their glycerides, which appear in the composition of saponifiable components.

Table 3.2. Basic fatty stock used in manufacture of greases

Stock	Standard	Grease type whose composition included this stock	Approximate consumption for greases, %
Cottonseed oil (CsO)	GOST 1128-75	Calcium, calcium-colium	8-11* 9-11*
Rapeseed oil (RO)	GOST 8988-77	as above	8-11* 9-11*
Sunflower oil (SO)	GOST 1129-73	as above	8-11* 9-11*
Castor oil (CO)	GOST 6757-73	Sodium, sodium-calcium *lithium*	18-20 18-20 4-6*
Ground animal fats: beef, lamb, pork ones (GAF)	—	Calcium, calcium-sodium	3-5* 4-6*
Sea animal fats: sperm whale, spermacetic ones (SAF)	—	Lithium	5-10*
Salomas (S)	OST 18-373-81	Calcium, calcium-sodium, sodium	3-5* 4-6* 2-22
Commercial stearic acid (HSt)	GOST 6484-64	Calcium complex, sodium lithium Barium complex, aluminium, Aluminium complex	10-15 5-10* 13-16 5-7* 10-12* 8-12
Commercial oleic acid (HOl)	GOST 7580-55	Lithium, aluminium	5-7* 10-15*
Hydrogenated castor oil (HCO)	STP 3-1-73	Lithium, anhydrous calcium	7-13 8-12
Commercial 12-oxystearic acid (12-HoSt)	STP 3-2-73	Calcium complex, anhydrous calcium, lithium, lithium complex	9-11 8-12 6-12 8-10

SFA C_{20} and more stillage residue	OST 38-01189-80	Calcium	7-9*
SFA C_5-C_6 and C_7-C_9	GOST 22239-78	Calcium	4-5
SFA C_{10}-C_{20} or mixture C_{10}-C_{17} + C_{17}-C_{20}	STP 07-19-76	Calcium complex, Aluminium complex	9-11 9-11
Non-thermally treated SFA	—	Calcium, sodium-calcium	13-17* 20-25
Oxidized petrolatum (OP)	—	Semiliquid calcium	4-5*
Naphthenic acids (NA)	GOST 13302-77	Semiliquid calcium	4-5*

* Fatty components used in mixture with another saponifiable stock.

The composition of natural animal and vegetable fats includes glycerides or high molecular weight esters (waxes) of monobasic straight-chain fatty acids or, as they are called, normal fatty acids from C_4 to C_{30} [8-14].

Table 3.3. Saturated fatty acids contained in natural fats

Acid	Chemical formula	Number of carbon atoms	Molecular mass	Acid number, mg KOH/g	Melting point, °C	Density*, kg/m³
Butanoic (butyric)	C_3H_7COOH	C_4	88.1	636.8	-7.9	959 (20)
Hexanoic (carpoic)	$C_5H_{11}COOH$	C_6	116.2	483.0	-3.4	929 (20)
Octanoic (caprylic)	$C_7H_{15}COOH$	C_8	144.2	389.0	16.7	910 (20)
Decanoic (capric)	$C_9H_{19}COOH$	C_{10}	172.3	325.7	31.6	895 (30)
Dodecanic (lauric)	$C_{11}H_{23}COOH$	C_{12}	200.3	280.1	44.2	883 (20)
Tetradecanoic (myristic)	$C_{13}H_{27}COOH$	C_{14}	228.4	245.7	53.9	851 (70)
Hexadecanoic (palmitic)	$C_{15}H_{31}COOH$	C_{16}	256.4	218.8	62.9	848 (70)
Octadecanoic (stearic)	$C_{17}H_{35}COOH$	C_{18}	284.5	197.2	69.6	846 (70)
Eicosoic (arachic)	$C_{19}H_{39}COOH$	C_{20}	312.6	179.5	75.3	824 (100)
Docosanoic (behenic)	$C_{21}H_{43}COOH$	C_{22}	340.6	164.7	79.9	822 (100)
Tetracosanoic (lidnoceric)	$C_{23}H_{47}COOH$	C_{24}	368.8	152.2	84.2	821 (100)
Hexacosanoic (cerotinic)	$C_{25}H_{51}COOH$	C_{26}	396.7	141.5	87.7	820 (100)
Octacosanoic (montanic)	$C_{27}H_{55}COOH$	C_{28}	424.7	132.1	90.0	819 (100)
Tricontanic (melissic)	$C_{29}H_{59}COOH$	C_{30}	452.8	123.9	93.6	819 (100)

*In brackets: the temperature (°C) at which the density was determined.

Saturated fatty acids are described by a general formula $C_nH_{2n+1}COOH$ or $C_nH_{2n}O_2$. Some characteristics of saturated fatty acids whose glycerides are contained in natural fats are given in Table 3.3 [8-12]. Of the saturated fatty acids, natural fats mostly contain C16 and C18. Normal saturated fatty acids appear primarily in the composition of animal fats, while in vegetable oils they are in small amounts.

Table 3.4. Unsaturated fatty acids contained in natural fats

Acid[1]	Chemical formula	Number of carbon atoms	Molecular mass	Acid number, mg KOH/g	Iodine number, mg I_2/100g	Melting point, °C	Density kg/m³
Oleic-series fatty acids							
Cis-9-Tetradesenic (myristoleic)	$C_{13}H_{25}COOH$	C^1_{14}	226.4	247.9	111.9	-4.5	902(20)
Cis-9-Hexadecenic (palmitoleic)	$C_{15}H_{29}COOH$	C^1_{16}	254.4	220.5	99.8	0.5	-
Cis-6-Octadecenic (petroselinic)	$C_{17}H_{33}COOH$	C^1_{18}	282.4	198.6	89.9	30.0	868(70)
Cis-9-Octadecenic (oleic)	$C_{17}H_{33}COOH$	C^1_{18}	282.4	198.6	89.9	13.4	890(20)
Trans-9-octadesenic (elaidic)	$C_{17}H_{33}COOH$	C^1_{18}	282.4	198.6	89.9	46.5	857(70)
Trans-11-Octadese-nic (vaccenic)	$C_{17}H_{33}COOH$	C^1_{18}	282.4	198.6	89.9	44.0	-
Cis-9-Eicosenoic (gadoleic)	$C_{19}H_{37}COOH$	C^1_{20}	310.5	180.7	81.7	25.0	-
Polyolefinic acids							
Cis-9, cis-12-Octa-decadienoic (linolic)	$C_{17}H_{31}COOH$	C^2_{18}	280.4	200.1	181.1	-5.0	-
Cis-9, cis-12, cis-15-Octadecatrienoic (linolenic)	$C_{17}H_{29}COOH$	C^3_{18}	278.4	201.5	273.5	-10.0	-

1. Numerals 6, 9, 11 etc. Indicate the No. Of unsaturated carbon atom which is closer to the carboxy 1 group.

2. Superscript indicates the number of double bonds in the molecule of the acid.

3. In brackets: the temperature (°C) at which the density was determined.

All vegetable oils contain glycerides of unsaturated fatty acids, some of which have two double bonds. The number of isomers of these acids is greater than that of monounsaturated acids. Most widespread acids with two isolated double bonds are cis-9- and cis-12-octadecadienoic (linoleic) acids.

Main characteristics of unsaturated fatty acids included in the composition of natural fats that are employed as saponifiable components of greases are given in Table 3.4 [8-11]. It should be noted that, apart from the acids listed in Table 3.4, the compositions of fats also include other isomers of unsaturated acids, but their contents are insignificant.

In addition to glycerides of saturated and unsaturated fatty acids, natural fats also contain glycerides of acids with functional groups in the radical: hydroxyl group (OH), for hydroxyacids; carbonyl group (CO), for ketoacids. A very valuable acid in the grease production is 12-hydroxyoctadecanoic (12-hydroxystearic) acid, obtained by hydrogenation of castor oil, whose composition includes glycerides of 12-hydroxy-cis-9-octadecenoc (ricinoleic) acid. We will dwell separately on the properties of this acid and method of its production, since the use of 12-HoSt or its glycerides, hydrogentated castor oil (HCO), is of particular importance in grease production. In the former USSR, 12-HoSt is commercially produced primarily at grease factories.

Natural fats are a scarce stock, and therefore some greases are produced in the former USSR with the use of synthetic fatty acids (SFA) obtained by oxidation of paraffins [14]. The SFA compositions contain mainly normal saturated fatty acids of even (Table 3.3) and odd (Table 3.5) number of carbon atoms, ranging from 4 to 28 [5,14]. They also contain iso-structure fatty acids (15-25%), dicarboxylic acids (4-5%), and small amounts of unsaturated acids, keto- and oxyacids, and unsaponifiable substances.

All the above-mentioned acids consisting of 10 and more carbon atoms exhibit weak acidic properties and practically do not dissociate. The measure of acidic properties of fatty acids or their mixtures is the acid (neutralization) number (AN), expressed by the amount of KOH (mg) needed to neutralize 1 g of the substance under investigation. The acid number and the molecular weight (M) of an acid are interrelated: $AN = 56.110:M$.

At room temperature, saturated fatty acids up to C9 are liquids, and C10 and above are solids. An important characteristic of fatty acids is their melting (congealing) point. From the thermodynamic point of view the concepts of melting point and congealing point are identical, but the congelation point of fatty acids is practically always 1-3 °C below their melting point. It should be pointed out that the melting point of fatty acids is very sensitive to slightest contaminations and there are discrepancies in numerical values of melting points of fatty acids presented in technical literature.

Crystallization of mixtures of fatty acids, employed in grease production proceeds in a more complex manner than that of individual ones, since the equilibrium in cooling of the melt of acid mixtures is reached very slowly: it sets in at a cooling rate not over 0.1-0.2 °C/day; mixtures of fatty acids are therefore practically always in a nonequilibrium state. When cooled, they form eutectic mixtures. The features of crystallization of fatty acids affect not only their properties, but also the characteristics of corresponding soaps. This in turn affects the structure of the skeletons of greases and their properties, which should be taken into account in the production of greases.

As the molecular weight of fatty acids increases, their density decreases and their boiling point rises. All the fatty acids, except the first two homologues, are lighter than water. A thermal treatment (240-260°C) of fatty acids results in their decarboxylation and formation of hydrocarbons or ketones

$$RCOOH \longrightarrow RH + CO_2,$$

$$2RCOOH \longrightarrow R-\overset{\overset{O}{\|}}{C}-R + CO_2 + H_2O.$$

Table 3.5. Saturated fatty acids contained in natural fats

Acid	Chemical formula	Number of carbon atoms	Molecular mass	Acid number, mg KOH/g	Melting point, °C	Density*, kg/m³
Methanoic (formic)	HCOOH	C_1	46.0	1219.0	8.4	1220(20)
Propanoic (propionic)	C_2H_5COOH	C_3	74.1	757.4	-22.0	992(20)
Pentanoic (valerianic)	C_4H_9COOH	C_5	102.1	549.3	-34.5	942(20)
Heptanoic (enanthic or heptylic)	$C_6H_{13}COOH$	C_7	130.2	431.0	-10.5	922(20)
Nonanoic (pelargonic)	$C_8H_{17}COOH$	C_9	158.2	354.6	12.5	907(20)
Hendecanoic (undecylic)	$C_{10}H_{21}COOH$	C_{11}	186.3	301.2	29.3	894(25)
Tridecanoic (tridecylic)	$C_{12}H_{25}COOH$	C_{13}	214.3	261.8	41.5	846(80)
Pentadecanoic (pentadecylic)	$C_{14}H_{29}COOH$	C_{15}	242.4	231.5	52.3	842(80)
Heptadecanoic (margaric)	$C_{16}H_{33}COOH$	C_{17}	270.4	207.5	61.3	853(60)
Nonadecanoic (nonadecylic)	$C_{18}H_{37}COOH$	C_{19}	298.5	188.0	68.6	877(24)
Heneicosanic	$C_{20}H_{41}COOH$	C_{21}	326.6	171.8	74.3	824(100)
Tricosanic	$C_{22}H_{45}COOH$	C_{23}	354.6	158.2	79.1	822(100)
Pentacosanic	$C_{24}H_{49}COOH$	C_{25}	382.7	146.6	83.5	—
Heptacosanic	$C_{26}H_{53}COOH$	C_{27}	410.7	136.6	87.6	820(100)
Nonacosanic	$C_{28}H_{57}COOH$	C_{29}	438.8	127.9	90.3	819(100)

*In brackets: the temperature (°C) at which the density was measured.

Acids (C_1 and C_2) dissolve in water in any ratio. As hydrocarbon chain increases, the water solubility of acids declines. Thus, 0.015 g of caprylic acid (C_{10}), but only 0.00029 g of stearic acid (C_{18}) dissolve in 100 g of water at 20°C. The water solubility of the acids increases with temperature (C_{18} at 60°C, 0.005 g in 100 g). Fatty acids are also able to dissolve water, this capability decreasing with rising molecular weight of the acids and increasing with temperature. Fatty acids dissolve better in water than in such solvents as acetone (C_{18}, 1.54 g in 100 g at 20°C), benzene (C_{18}, 2.5 g in 100 g at 20°C), alcohols (ethanol: C_{18}, 0.65 g in 100 g at 20°C), and other polar and nonpolar solvents. Fatty acids C10-C14 dissolve well in petroleum oils. Higher molecular weight acids (C_{16}-C_{20}) dissolve petroleum oils on heating. Unsaturated fatty acids dissolve in organic solvents somewhat better than saturated ones and facilitate the dissolution of the latter. Fatty acids are capable of chemical reactions with participation of both the carboxylic group and the hydrocarbon radical.

The reaction with participation of the carboxylic group, i.e., formation of salts (soaps), is important in the grease production process.

Alkali metal soaps of fatty acids hydrolyse in presence of water, and therefore this reaction is reversible for them. Reacting with alkaline-earth metal hydroxides, fatty acids can form basic and neutral salts, which should be taken into account in production of Ca- and Ba-greases. Fatty acids also form acid salts.

Metallic soaps, salts of heavier metals, are produced by double-exchange reactions i.e., as a result of interaction of alkali metal soaps with water-soluble salts of heavy metals. At an elevated temperature, metallic soaps can also result from a direct interaction of fatty acids with some metals, such as iron. This explains the corroding action of fatty acids on carbon-steel production equipment. The corroding action of fatty acids intensifies as the molecular weight of the acids decreases and the temperature increases, which is to be taken into consideration in selecting the grease production equipment.

Stable soaps of trivalent metals, such as aluminium, as well as of higher valency metals cannot be produced by the above methods. Aluminium soaps produced by precipitation from an aqueous solution of sodium soap by aluminium sulfate solution have various polymeric structures, mixtures of mono-, di-, and trisubstituted soaps, whose composition, structure, and properties depend on many factors. More stable aluminium soaps result from interaction of alkoxyaluminium with fatty acids in an anhydrous organic solvent. Aluminium soaps can be described by a general formula $(RCOO)_m Al(OH)_n$, where $m + n = 3$. Most suitable for production of greases is a disubstituted soap, i.e., one where $m = 2$ and $n = 1$.

An important chemical reaction of fatty acids with participation of the carboxylic group is their interaction with various alcohols, yielding esters:

$$RCOOH + R_1OH \rightleftharpoons RCOOR_1 + H_2O.$$

Esters resulting from interaction of various acids and alcohols are used as dispersion media of greases.

Esters of fatty acids and tribasic, glycerol, a trihydric alcohol is a component of any fat; they are called glycerides. Fats contain primarily triglycerides, i.e., those formed by glycerol and three acid molecules. Triglycerides can contain radicals of either one fatty acid (single-acid triglycerides) or different acids (different-acid ones). Natural fats usually contain different acid glycerides. Physical properties of glycerides vary with their composition and **crystal** structure. An important reaction of glycerides is hydrolysis, resulting in glycerol and **free fatty** acids:

$$\begin{array}{c} CH_2OCOR \\ | \\ CH_2OCOR \\ | \\ CH_2OCOR \\ \text{glyceride} \end{array} + 3H_2O \rightleftharpoons \begin{array}{c} CH_2OH \\ | \\ CH_2OH \\ | \\ CH_2OH \\ \text{glycerol} \end{array} + 3RCOOH \quad \text{acid}$$

The splitting of glycerides is also effected by action of caustic alkalis on them:

$$\begin{array}{l} CH_2OCOR \\ | \\ CHOCOR \\ | \\ CH_2OCOR \end{array} + 3MOH \rightleftharpoons \begin{array}{l} CH_2OH \\ | \\ CH_2OH \\ | \\ CH_2OH \end{array} + 3RCOOMe$$

This reaction takes place in production of greases with the use of natural fats. Glycerol remains in the grease in this case; featuring pronounced properties of a surfactant, it substantially affects the formation of the skeleton and the properties of greases [15].

To produce glycerol-free fatty acids, the formed soap is decomposed by a mineral (most often, hydrochloric) acid. For this purpose the reaction of interaction of glycerides with alkalis and then that of the formed soaps with mineral acids are conducted in an aqueous medium. Water-insoluble fatty acids are separated out, washed, dried, and used for grease production; commercial glycerol is isolated from the aqueous solution.

Reaction involving the hydrocarbon radical of fatty acids are of minor importance in grease production and will therefore not be dealt upon. Note, however, that unsaturated fatty acids are most reactive in this respect, and therefore greases produced with soaps of these acids or of their glycerides exhibit a low chemical stability.

Let us discuss now the methods of production and characteristics of the saponifiable stock (Table 3.2) widely used in grease production.

Vegetable oils [8-10]. Cottonseed and castor oils and, less often, sunflower and rapeseed oils are used in grease production in the erstwhile USSR. Abroad, apart from these oils, palm, coconut and some other oils are employed. Vegetable oils are separated out from fruits and seeds of oil-bearing crops (sunflower, castor bean plant, rape, palm fruits, etc.) by expression, solvent extraction, or a combined method, *i.e.*, expression and extraction. Depending on the degree of refining, the oils are divided into refined and nonrefined ones.

Characteristics of vegetable oils used as the saponifiable stock for grease are presented in Table 3.6 [8-10].

Cottonseed, sunflower and rapeseed oils are primarily used in production of hydrated Ca-greases (fatty cup greases) and calcium-sodium greases, such as grease IP-1. Castor oil is employed to produce sodium, sodium-calcium, and lithium greases and also serves as the starting stock to produce 12-hydroxystearic acid, which is very valuable in the grease production.

Ground animal fats [9-12]—beef, lamb, and pork ones - are seldom used to produce greases. They are primarily employed to produce hydrated calcium greases. Animal fats are as a rule separated out by melting. Their characteristics are presented in Table 3.7 [9-12]. They contain glycerides of the same fatty acids: myristic (C_{14}), palmitic (C_{16}), stearic (C_{18}), oleic (C_{18}^1), and linolec (C_{18}^2) ones.

Sea animal fats [10-13]—Sperm whale and spermacetic fats are most widely used in the grease production. They are obtained in the processing of sperm whales: sperm whale fat, from the body of the animal, and spermacetic fat, from bony skull recesses of its head. They are mainly used in a sulphurised form in production of lithium and other greases. Sperm whale and spermacetic fats contain 60-75% waxes and 25-40% glycerides; the spermacetic fat has an especially high content of waxes (up to 75%). The waxes contain mainly saturated cetyl (C_{16}) and stearyl (C_{18}), unsaturated palmitoleyl (C_{16}^1) and oleyl (C_{18}^1) alcohols. The fatty-acid composition of the fats and their other properties are presented in Table 3.8 [10-13].

Products of hydrogenation of vegetable oils, animal and fish fats are in the former USSR called **salomases** (hydrogenated fats) [9-12]. They differ in the fatty-acid composition, titre (congealing point), iodine number, and other characteristics depending on the starting stock

Table 3.6. Characteristics of vegetable oils used as saponifiable stock for grease

Characteristics	Oil			
	Cottonseed (GOST 1128-75)	Rapeseed (GOST 8988-77)	Sunflower (GOST 1129-73)	Castor (GOST 6757-73)
Iodine number, mg I_2/100 g	116-125	94-106	125-145	82-91
Acid number, mg KOH/g	0.2-14.0	0.4-6.0	2.2-6.0	1.6-5.0
Saponification number, mg KOH/g	189-199	165-180	186-194	176-186
Density at 20°C, kg/m^3	918-932	908-915	917-920	947-970
Flash point, °C (not below)	225	230	225	240
Pour point, °C	From 5 to −6	From 0 to −10	From −15 to −19	Not over −16
Content of acids, mass. %:				
palmitic	17.9-22.5	1.0-3.0	3.5-6.4	0.8-2.0
stearic	1.6-4.9	0.2-3.0	1.6-4.6	0.7-2.1
atachic	1.1-1.2	—	0.7-0.9	0.2-0.8
behenic	—	0.6-2.5	—	—
oleic	16.6-26.6	15.0-32.0	25.0-35.0	2.0-5.2
linolic	45.0-59.4	13.0-25.0	55.0-72.0	3.4-7.8
linolenic	—	7.0-10.0	—	0.2-0.8
eicosenoic	—	8.0-15.0	—	—
erucic	—	40.0-54.0	—	—
ricinolic	—	—	—	80.0-90.2

used and its hydrogenation conditions. Salomases contain primarily glycerides of saturated and unsaturated fatty acids. Unsaturated acids include starting ones, which have not undergone hydrogenation, and newly formed ones, resulting from a partial saturation of polyunsaturated acids. The latter acids are called isooleic; their molecules have a double bond whose position differs from that of the double bond in oleic acid. Isooleic acids are solids; their melting points are within 30-50°C. Therefore at the same titre (46-48°C) the iodine number of a salomas produced from vegetable oils (especially linseed and sunflower oils) as well as from fish fats (65-75) is much higher than that of a salomas produced from animal fats (40-45).

---

*The superscript numeral indicates the number of double bonds in the molecule of an unsaturated acid or alcohol.

These features of salomas should be taken into account in grease production. Main characteristics of salomases are given in Table 3.9 [9-12].

Table 3.7. Characteristics of ground animal facts

Characteristics	Beef fat	Lamb fat	Pork fat	Bone (beef) fat
Temperature, °C, of:				
congelation of fat	34-38	34-45	22-32	34-38
melting of fat	42-52	44-55	28-48	36-34
melting of fatty acids	40-77	39-45	27-40	—
Density at 15°C, kg/m^3	937-953	937-961	915-923	931-938
Relative content of acids, mass. %:				
saturated	52-60	52-62	37-46	39-42
unsaturated	40-48	38-48	54-63	58-61
Saponification number, mg KOH/g	193-200	191-206	193-200	193-198
Iodine number, g I$_2$/100g	32-47	35-46	46-66	43-60

Table 3.8. Characteristics of sperm whale fats

Characteristics and composition		Sperm whale fat	Spermacetic fat
Temperature, °C, of:			
congelation		7-15.5	40-45
melting		Up to 18	42-47
Density at 15°C, kg/m^3		873-890	945-970
Saponification number, mg KOH/g		112-163	112-163
Iodine number, g I$_2$/100 g		62-123	62-123
Content of fatty acids (alcohols), %:		(–) (–)	
saturated			
capric (decyl)	— C$_{10}$	(–) (–)	3.5 (–)
lauric (lauryl)	— C$_{12}$	1.0 (–)	16.0 (–)
myristic (myristyl)	— C$_{14}$	5.0 (traces)	14.0 (2-5)
palmitic (cetyl)	— C$_{16}$	6.5 (10-25)	8.0 (30-36)
stearic (stearyl)	— C$_{18}$	(–) (5-14)	2.0 (12-15)
unsaturated			
lauroleic	— C$_{12}$	(–) (–)	4.0 (–)
myristoleic (tetradecyl)	— C$_{14}$	4.0 (–)	14.0 (up to 5)
palmitoleic (hexadecyl)	— C$_{15}$	26.5 (7-13)	15.0 (10-17)
oleic (octadecyl)	— C$_{18}$	37.0 (40-53)	17.0 (32-36)
gadoleic (eicosenyl)	— C$_{20}$	19.0 (6-8)	6.5 (1-10)
clupanodonic (docosenyl)	— C$_{22}$	Up to 1.0 (traces)	Up to 1.0 (up to 2)

Table 3.9. Characteristics of commercial salomas (OST 18-373-81)

Salomas	Iodine number, mg $I_2/100\ g$ (not over)	Acid number, mg KOH/g (not over)	Congelation point, °C
For toilet soap (type 3):			
from oils and fats	65.0	3.5	39-43
from distilled fatty acids	65.0	Not specified	39-43
For laundry soap (type 4):			
from oils and fats	65.0	5.0	46-50
from distilled fatty acids	55.0	Not specified	46-50
For stearine (type 5), grades:			
5-1	2.5	8.0	Not below 65
5-2	9.0	8.0	59
5-3	17.0	7.0	58
5-4	30.0	6.0	53

Salomases mixed with the cottonseed or other vegetable oil are more often used to produce calcium greases and in limited amounts as the saponifiable stock in production of sodium and lithium greases.

Fatty acids [9-12]. Commercial normal saturated (stearic) and unsaturated (oleic) acids are widely used in grease production [9-12]. They are separated out from vegetable oils, animal fats, and salomases by splitting with water at 220-250°C and a pressure of $(25-60) \cdot 10^5$ Pa. The resulting mixture is distilled to remove accompanying substances, and then fatty acids are separated into solid and liquid ones by crystallization or expression. This yields two products: solid press-stearin with a titre of 40-50°C and a mixture of liquid acids (oleomargarine) with a titer of 12-25°C. The composition and properties of these products depend on the starting stock composition. Solid press-stearine is used to produce commercial stearins and stearic acid of various purity grades; their characteristics are presented in Table 3.10 [9-12]. Commercial stearin of the first and second grades is mainly used for production of greases, while that of the special grade is used in some cases. Commercial stearin serves as the saponifiable stock in production of lithium and, less often, aluminium, barium, and sodium greases.

The best stock to produce oleic acid is olive oil, which contains up to 85 percent triglycerides of this acid. A mixture of corriander (70%) and mustardseed (30%) or rapeseed (30%) oils was used in the former USSR. Soap stocks of the cottonseed oil and, in very limited amounts, sperm whale and liquid pork fats are also used.

Commercial oleic acid, whose characteristics are given in Table 3.11, is employed in production of some lithium, aluminium and other greases. Grade B is more often used; it contains 4-5 % C_{26}, 1-2 % C_{18}, 20-25 % C_{18} with a double bond in the [6-7] position, 17-23 % C_{18} with a double bond in the [9-10] position, 22-24 % C_{18}^2, 10-14 % C_{18}^3, and 11-15 % C_{20} acids [12].

Table 3.10. Characteristics of commercial (stearine) and reactive stearic acids

Acid	Iodine number, mg I$_2$/100 g (not over)	Acid number, mg KOH/g	Congelation point, °C (not below)	Mass fraction of stearic acid ($C_{18}H_{36}O_2$)
Commercial stearic acid (stearine), GOST 6484-64:				
special grade A	3.0	198-201	65.0	Depends on starting stock composition
B	10.0	194-210	59.0	"
I	18.0	198-210	58.0	"
II	32.0	198-210	53.0	"
Reactive stearic acid, GOST 9419-78:				
analytically pure	2.0	197-198	69.0	At least 99.0%
pure	3.0	198-200	67.0	At least 98.0%

Table 3.11. Characteristics of commercial oleic acid (GOST 7580-55)

	Distilled		Undistilled
	grade A	grade B	grade V
Color	yellow to light-brown	yellow to light-brown	yellow to dark-brown
Content of fatty acids in terms of oleic acid, % (not below)	95	95	92
Content of unsaponifiables, % (not over)	3.5	3.5	6.5
Congelation temperature, °C (not below)	10	16	34
Iodine number, mg I$_2$/100 g	80-90	80-90	Not specified
Acid number, mg KOH/g	185-200	185-200	At least 175
Transparency at 70°C	Transparent	—	Opaque

12-hydroxystearic acid [12,16]. The source for its production is castor oil (see Table 3.6), which contains up to 85 % ricinolec (12-hydroxyoleic) acid in the form of triglycerides.

To produce 12-HoSt, castor oil is subjected to hydrogenation, with the result that ricinolec acid gets saturated at the place of the double bond and transforms into 12-hydroxystearic acid.

DISPERSED PHASE OF LUBRICATING GREASES

$$CH_3-(CH_2)_5-\overset{\overset{\displaystyle OH}{|}}{CH}-CH_2-CH=CH-(CH_2)_7-COOH + H_2$$
(ricinolec acid —HRi)

$$\longrightarrow CH_3-(CH_2)_5-\overset{\overset{\displaystyle OH}{|}}{CH}-(CH_2)_{10}-COOH$$
(12-hydroxystearic acid)

In the hydrogenation process, castor oil transforms into a solid mass, hydrogenated castor oil (HCO), consisting of glycerides of 12-hydroxystearic (up to 80 %), ricinolec (2-4%), stearic with traces of palmitic (10-15 %), and oleic with traces of linolec (1-4 %) acids. 12-hydroxystearic acid is isolated from HCO by saponifying the latter with a 20-25-% NaOH solution, followed by decomposing the soaps obtained with hydrochloric acid.

Characteristics of HCO and commercial 12-HoSt, used for the grease production, are presented in Table 3.12 [5, I6].

Table 3.12. Characteristics of hydrogenated castor oil and 12-hydroxy stearic acid

Characteristics	Hydrogenation castor oil (TU 38.101726-78)	12-hydroxy stearic acid (TU 38.101721-78)
Appearance	White, with light-blue shade, solid brittle product	White solid product
Melting point, °C	Not below 84	74-76
Acid number, mg KOH/g	Not over 2	170-180
Saponification number, mg KOH/g	175-185	180-190
Iodine number, mg I_2/100 g	5-10	5-10

Content, %:
   oxyacids (not below)
   ash (not over)
   water
   mechanical impurities
	
	 82
	 0.02
	 Absent
	 Absent
		
		 82
		 0.02
		 —
		 —

Hydrogenated castor oil (TU 38 101726-78) and commercial 12-hydoxystearic acid (TU 38 101721-78) are employed to produce lithium, anhydrous calcium, and other greases. A HCO—12-HoSt mixture can also be used [15]. Properties of hydrogenated castor oil and of 12-hydroxystearic acid isolated from it depend substantially on the quality of castor oil, conditions of its hydrogenation, saponification of hydrogenated castor oil, decomposition of soaps by hydrochloric acid, washing, drying, and storage of commercial 12-HoSt.

Refined first-grade castor oil should be used for the hydrogenation and the process is to be conducted in presence of medium-activity selective nickel catalysts at 150-170°C and a pressure of $(8\text{-}25) \cdot 10^5$ Pa. It should be pointed out that the castor oil hydrogenation, apart from the main reaction of ricinolic acid saturation at double bonds, can involve dehydration of the latter with formation of linolec acid, followed by its transformation into stearic acid. This reaction occurs at elevated (over 180°C) temperatures. Moreover, at a temperature above 200°C splitting of glycerides and then a hydrogenolysis reaction can occur, the latter resulting in transformation of acids into alcohols and neutral hydrocarbons; this reduces the saponification number of the finished product and increases the content of unsaponifiables in it. With lack of hydrogen, the reaction of dehydration of 12-hydroxystearic acid with the formation of 12-ketostearic acid takes place, while in an excess of hydrogen, an elevated temperature, and a deep hydrogenation (iodine value below 2 g I_2/100 g), 12-hydroxystearic acid transforms into stearic acid. The presence of a hydroxyl and a carboxyl group in 12-hydroxystearic acid favours the formation of internal and intermolecular esters. Experiments ascertained (Fig. 3.1) [17] that storage of 12-hydroxystearic acid in a molten state at elevated temperatures results in a drastic decrease of its acid number, which not only has an adverse effect on the grease production, increasing the consumption of the acid, but also significantly impairs the quality characteristics of the grease. All this should be taken into account in the production, storage, and use of 12-hydroxystearic acid, where the production process regulations must be strictly observed.

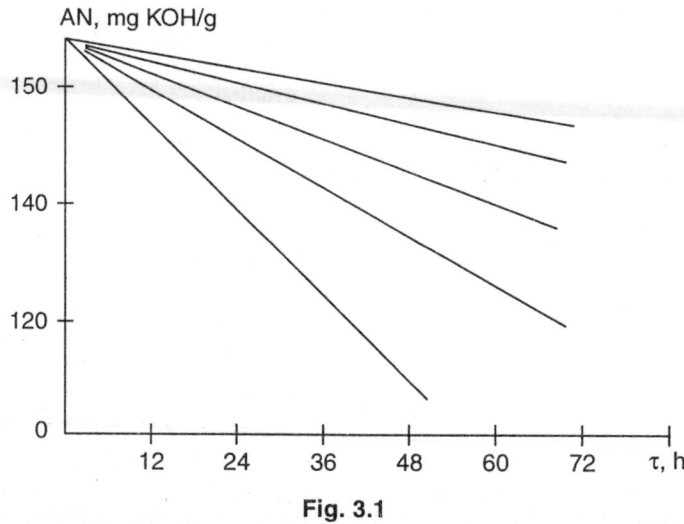

Fig. 3.1

Synthetic fatty acids [12-14]. The starting stock for producing SFA is paraffin. It is oxidized by the oxygen in air in presence of catalysts. The resulting oxidized paraffin is subjected to saponification by a caustic alkali solution. Unoxidized paraffin and oxygen-containing compounds neutral with respect to alkalis (unsaponifiables) are separated from soaps by the method of settling in aqueous solutions and thermal treatment and are then returned for a repeated oxidation. Soaps are decomposed by acids (H_2SO_4 or HNO_3), and a broad SEA fraction, consisting of acids from C_5 to C_{20} and above is separated out. This broad SFA fraction is separated by a vacuum distillation into narrow ones: C_5-C_6, C_7-C_9, C_{10}-C_{16}, C_{17}-C_{20}, and C_{20} and above acid fractions. The acids are used for various technical purposes, including grease production. Such acids as C_{20} and above (stillage residues) in a mixture with C_5-C_6 (C_7-C_9) acids are employed in production of hydrated calcium greases and synthetic cup greases. Synthetic fatty acids C_{10}-C_{20} (C_{10}-C_{16} + C_{17}—C_{20}) are widely used in production of cCa- and cAl-greases. Attempts were undertaken to use C_{10}-C_{16} SFA to prepare lithium greases, but no positive results were obtained.

Apart from the above-mentioned thermally treated and distilled SFA, the use in production of some greases (hydrated calcium and sodium-calcium greases) is found of a thermally untreated SFA fraction, which apart from C_5-C_{20} acids, contains also some unsaponifiables, i.e., those that are separated during the thermal treatment of soaps. The acid number (AN) of these acids is usually of 100-120 mg KOH/g; and the saponification number (SN) 160-170 mg KOH/g. Synthetic fatty acids C_{20} and above and stillage residues are commercially produced to OST 38-01189-80 and have the following characteristics: AN = 100-125 mg KOH/g; SN = 140-165 mg KOH/g. A stillage residue contains about 50% polymeric compounds, about 35% C_{28}-C_{40} SFA, and 10-15% unsaponifiables.

Characteristics of thermally treated and distilled SFA, commercially produced according to their effective standards and specifications, are presented in Table 3.13. However, the specified characteristics incompletely characterize the properties of commercial SFA fractions. Essential for grease production is their composition, which does not always coincide with the commercial marking. Thus, the C_{10}-C_{16} SFA fraction contains 90-95% C_{10}-C_{18}, 4-6% C_7-C_9, and 3-4% C_{19}-C_{20} acids. The higher the molecular weight of the acids, the lower is their degree of purity. Apart from fatty acids with even and odd numbers of carbon atoms, SFA also contain iso- oxy- and ketoacids and other oxygen-containing organic compounds, whose percentage is insignificant. They primarily (75-90 %) contain straight chain saturated fatty acids (Table 3.14). More narrow acid fractions, which can as well be used in the grease production, are obtained from C_{10}-C_{16} and C_{17}-C_{20} SFA by a repeated distillation process.

Oxidized petrolactum is sometimes used to produce greases. It is oxidized in the same plants as is paraffin. Petrolatum is much more difficult to oxidize than paraffin. It is used in an oxidized state without additional treatment and sometimes in a sulphurised state. Oxidized petrolatum is characterized (TU 38. 30196-78) by the acid number and saponification number (AN = 45-50 mg KOH/g; SN = 150-160 mg KOH/g). It is nontoxic at normal temperatures; the maximum permissible concentration of evolved hydrocarbon vapours in the working zone air is 300 mg/m3; the flash point is 240°C.

Table 3.13. Characteristics of commercial thermally treated and distilled SFA fractions (GOST 23239-78)

Commercial fractions of acids	Acid number, mg KOH/g	Esterification number, mg KOH/g (not over)	Carboxylic number, mg KOH/g (not over)	Mass fraction of unsaponifiables, % (not over)	Mass fraction of water, % (not over)	Congelation temperature, °C	Mass fractions of acids, %						
							C_5-C_6 (at least)	Up to C_7 (not over)	C_5-C_6 (at least)	Up to C_{10} (not over)	Above C_9 (not over)	C_{10}-C_{20} (at least)	Above C_{20} (not over)
C_5-C_6 (State Quality Mark)	480-550	—	—	0.5	4.0	—	70.0	—	—	—	—	—	—
C_7-C_9 (State Quality Mark)	370-410	—	—	0.8	1.5	—	—	15.0	80.0	—	10.0	—	—
C_7-C_9 -First grade	370-410	—	—	1.0	1.5	—	—	17.0	75.0	—	11.0	—	—
-Second grade	370-410	—	—	1.0	1.5	—	—	17.0	72.0	—	11.0	—	—

$C_{10}-C_{16}$ (State Quality Mark)	240-260	4.5	11.0	1.9	1.0	25-32	—	—	2.5	—	—	3.5
$C_{10}-C_{16}$	240-260	4.5	12.0	2.1	1.0	25-35	—	—	4.5	—	—	5.0
$C_{17}-C_{20}$, brand M: (State Quality Mark)	200-210	6.0	13.0	4.0	1.0	45-49	—	—	1.0	—	—	25.0
-First grade	195-210	6.5	14.5	4.4	1.0	45-51	—	—	—	—	—	—
-Second grade	195-210	6.5	14.5	4.5	1.0	45-51	—	—	1.0	—	—	35.0
$C_{17}-C_{20}$, brand R	195-210	6.5	14.5	4.5	0.3	45-51	—	—	1.0	—	—	25.0
$C_{10}-C_{20}$*	210-240	7.0	-	5.0	1.0	Not below 30	—	—	8.0	—	80.0	12.0

*Produced to STR 07-1976.

Table 3.14. Characteristics of commercial SFA fraction

Content of individual acids, % (chromatographic analysis)	C_5-C_6	C_7-C_9	C_{10}-C_{16}	C_{17}-C_{20}
C_5	17.1	8.3	—	—
C_6	26.2	17.1	—	—
C_7	22.6	25.0	0.7	—
C_8	18.4	23.0	1.0	—
C_9	9.8	16.5	4.0	—
C_{10}	5.8	7.7	7.9	0.3
C_{11}	—	2.5	10.4	0.7
C_{12}	—	—	10.1	1.9
C_{13}	—	—	11.2	2.9
C_{14}	—	—	12.2	5.3
C_{15}	—	—	11.4	6.8
C_{16}	—	—	10.9	8.3
C_{17}	—	—	9.6	9.9
C_{18}	—	—	7.0	12.2
C_{19}	—	—	2.3	11.9
C_{20}	—	—	1.3	11.8
C_{21}	—	—	—	11.5
C_{22}	—	—	—	9.5
C_{23}	—	—	—	4.7
C_{24}	—	—	—	2.3
Main fraction (purity over 85%)	C_5-C_8	C_6-C_9	C_{10}-C_{18}	C_{15}-C_{23}

Naphthenic acids: acidols and naphtha soaps [5, 12, 18].

Production of some greases involves the use of organic acids contained in distillate fractions of petroleums: kerosene, solar, spindle and machine oils. The acids vary in the composition and structure: they include normal acids, isoacids, naphthenic acids, aromatic acids, etc. It would therefore be more correct to call them petroleum acids, but their main components are naphtenic acids, which are mostly monobasic saturated acids containing five- and six-member carbon rings, of a general formula $C_nH_{2n-2}O_2$; higher molecular weight petroleum acids (of a general formula $C_nH_{2n-4}O_2$) include double and possibly also triple rings. Compositions of the acids depend on the petroleum field, petroleum processing method, petroleum fraction, from which they are separated out and on separation method.

Four groups of petroleum acids are produced on a commercial scale: distilled naphthenic acids; acidols, nondeoiled commercial naphthenic acids (two grades); acidols-naphtha soaps (two grades); and naphtha soaps (two grades) (Table 3.15).

Table 3.15. Characteristics of commercial petroleum acids (GOST 13302-77)

Characteristics	Distilled naphthenic acids*	Acidol A-1	Acidol A-2	Acidol-naphtene soap 1st grade	Acidol-naphtene soap 2nd grade	Nathene soap 1st grade	Nathene soap 2nd grade
Appearance	Transparent yellow liquid	—	—	Salvelike substance, from light-yellow to light-brown color			
Content,%:							
• petroleum acids (not bellow)	95	42	50	43	43	70	70
• unsaponifiables referred to inorganic part (not over)	3	57	45	9	13	9	13
• mineral salts (not over)	—	—	—	1	1	4	4
• water (not over)	—	5	3	—	—	—	—
Acid number of petroleum acids, mg KOH/g:							
• not over	220-260	175	210	—	—	—	—
• not below	—	—	—	220	210	220	210

Note. *Color of distilled naphthenic acids by iodine scale: not over 27.

Acidols are isolated from alkaline wastes of refining of distillate oils boiling within 300-420°C. The alkaline wastes are boiled down and salted out, and then decomposed by a mineral acid. Acidols-naphtha soaps are obtained from alkaline wastes of refining of kerosene and straw distillates, boiling within 180-300°C, after a partial neutralization of the wastes by 35-40% sulfuric acid. They contain about 70% free acids and about 30% naphthenic acid salts. Apart from petroleum acids, acidols-naphtha soaps contain a considerable amount of unsaponifiable substances. Acidols or acidols-naphtha soaps are more often used in production of greases or their components (lead and copper naphthenates), but their use is very insignificant.

Colophony and tall oil are occasionally used for the grease production as saponifiables and also simply as additives.

Colophony [5] is a brittle glasslike semi-transparent mass, from light-yellow to dark-brown in colour, featuring a conchoidal fracture. It consist of a mixture of resin (abietic) unsaturated acids (C_{19}—C_{29}) fatty acids (5-10%), and unsaponifiables (5-12%) and the following physico-chemical properties: density at 20°C, 1000-1085 kg/m^3; acid number, 150-175 mg KOH/g; softening point, 52-70°C; boiling point (at 665 Pa), 250°C; it dissolves well in ether, acetone, benzene, turpentine and gasoline. Basic characteristics of commercial colophony grades are presented in Table 3.16 [5].

Table 3.16. Characteristics of colophony of various grades

Characteristics	Pine colophony (GOST 19113-73)			Tallow colophony (GOST 14201-83)		
	A	B	K	A	B	V
Appearance	Transparent glasslike mass or mass containing air bubbles		Mass or mass	Transparent glasslike mass		
Coloring by color reference scale	X,WW, WG	N,M,K, L,H.G	F,E,D X,WW	Ng,N, M,K	J,H,G	F,E,D
Content, %:						
• Moisture (max)	0.3	0.3	0.4 0.2	0.2	0.2	0.2
• ash (max)	0.04	0.04	0.07 0.025	0.04	0.07	2.0
• mechanical impurities (max)	0.05	0.05	0.1 0.03	0.05	0.08	0.3
• unsaponifiables (max)	6.5	7.5	10.5 6.1	6.0	7.0	10.0
• volatiles A (max)	Not specified	Not specified	- 0.2	-	-	-
Softening point, °C	68	66	54 73	60	54	58
Acid number, mg KOH/g (min)	168	166	150 174	160	150	140

Note. Colophony of grade V is partly saponified.

Tall oil [5] is a liquid resinous mass of a dark colour, obtained from a by-product of sulfate pulp production, from sulfate soap by decomposing it with sulfuric acid. Tall oil contains approximately equal amounts of fatty acids (palmitic, stearic, oleic, linolic, etc.) and resin (mixture of abietic and oxidized resin) acids. It also contains up to 12% unsaponifiables, including up to 3% phytosterol. Unrefined tall oil has a pungent smell and a dark colour; distilled one is more bright and has no pungent odour. Distilled tall oil (TU 81-05-26-75) (Table 3.17) and tall fatty acids of grade B (GOST 14845-79) are used in grease production.

Table 3.17. **Characteristics of distilled tall oil (TU 81-05-26-75)**

Characteristics	First grade	Second grade	Third grade
Appearance	Brown oily liquid, transparent at 80°C	Brown oily liquid, transparent at 80°C	Brown oily liquid, transparent at 80°C
Acid number, mg KOH/g (min)	170	165	150
Saponification number, mg KOH/g (min)	175	170	150
Mass fraction of unsaponifiables, % (min)	6	9	13
Mass fraction of water, % (max)	traces	1	2
Flash point, °C (min)	200	200	200
Self-ignition point, °C (min)	327	327	327

3.1.3. Soaps

Soaps of alkali and alkaline-earth metals, the dispersed phase of greases, are mostly obtained in the grease production process (in situ). Finished air-dry soaps are used occasionally. Finished soaps are as a rule employed in production of metal (lead) soap base greases.

To produce alkali and alkaline earth metal soaps in situ, a part (generally, 1/3 of the calculated amount) of dispersion medium is charged into the kettle, and then saponifiable components (fats, fatty acids or petroleum acids) are melted in it at a temperature 5-10°C above their melting point. Next, an aqueous solution (suspension) of hydroxide of the corresponding metal is metered into the kettle in a thin stream. At 95-100°C, as a result of saponification of fats,

$$C_3H_5-(RCOO)_3 + 3 \text{ MeOH} \rightarrow 3RCOOMe + C_3H_5(OH)_3,$$

or neutralization of acids,

$$RCOOH + MeOH \rightarrow RCOOMe + H_2O,$$

followed by removal of the excess water from the reaction medium at 100-120°C, a soap concentrate in oil is formed, used for the subsequent production of lithium greases. They are produced in special plants by neutralisation of stearic or 12-hydroxystearic acid with a 10% aqueous solution of LiOH, followed by filtration, washing and drying of the formed soap. Finished soaps are also employed to prepare greases in laboratories for various investigations. Lithium, sodium and potassium soaps are produced (19) by adding an aqueous or an alcoholic solution of hydroxide of the corresponding metal to a molten starting acid under stirring. Neutralisation is carried out for an hour at 80-90°C. The precipitated lithium soap is filtered, washed with warm distillated water and dried under vacuum. When producing potassium and sodium soaps, the aqueous or aqueous-alcoholic solution is boiled down and the finished soap is dried; the spray drying, which provides a fine powder of air-dry soaps, can be used here. Finished dry soaps of alkaline earth metals are produced through a double-exchange reaction

(12, 19, 20) by precipitating them from a 15% aqueous or alcoholic solution of Na soap by chloride of the corresponding alkaline earth metal, using a 10% aqueous solution of the chloride with an excess of 5% over the theoritacally required amount. The reaction is continued under stirring for 1 hour. The soap, ion soluble in water, precipitates as a fine powder, which is then filtered, washed with warm distilled water and dried under vacuum. However, contamination of calcium or barium soaps with sodium ions may occur in this case. The saponification of fats and neutralization of acids are conducted with an excess of alkali, since lack of excess base/pH results in acidic soaps. The soap formation in situ involves a strong foaming, and therefore the temperature of the process should be monitored carefully and the kettle volume should be 3-4 times the volume of the soap concentrate produced. A particularly vigorous foaming occurs in production of calcium and sodium soaps.

For the production of calcium soap (20) fatty acids are heated in a water suspension of calcium hydroxide (milk of lime) which is taken with an excess of 1-2% over the theoretical amount. As the melting point of the acid is approached (2-3°C below it), the heating is stopped, while stirring the reaction mixture is continued; soap forms at this temperature. After the completion of the reaction, which is judged from retainment of weak pink colour of phenolphthalein, the mass is heated to boiling and boiled for 30 min. Next, it is washed with boiling distilled water till the aqueous extract becomes neutral, after which the soap is extracted and dried.

The authors of [21] proposed a method for production of dry calcium soaps, where fatty acids are charged into a vessel equipped with a vigorously stirring device and a heating device and heated to 125°C, after which, at a continuous vigorous stirring, the calculated amount of slaked lime is batch-metered into the vessel. When slaked lime is fed into the vessel, the temperature of acids being processed is raised at a rate of 1-1.5°C/min to 170-180°C, at that temperature the soap formation reaction is completed in 25-30 min. Next, the soap is cooled down to 40-45°C, discharged, and reduced to powder. Barium and alkali metal soaps can also be produced by this method.

Aluminium soaps are commercially produced by a double-exchange reaction [12, 22], treating an aqueous solution of Na-soap with a 30-35% aqueous solution of potassium alum taken with an excess of 1-2% over the theoretical amount:

$$6C_{17}H_{35}COONa + 2K[Al(SO_4)_2] \cdot 12H_2O \rightarrow 2(C_{17}H_{35}COO)_3Al + 3Na_2SO_4 + 24H_2O.$$

The aqueous solution of potassium alum is metered in by a thin stream at 50-60°C while the reaction mixture is continuously stirred. Next, the soap is separated from the mother liquor and washed with hot (50-60°C) water till the absence of SO_4 ions in the aqueous extract. The produced soap contains about 60% water and is in this form used for production of greases. Aluminium oleate is produced by this method. The reaction yields mono-($C_{17}H_{35}COOAl(OH)_2$), di-(($C_{17}H_{35}COO)_2AlOH$), and trisubstituted (($C_{17}H_{35}COO)_3Al$) aluminium soaps. Chemically most stable are mono-substituted, while trisubstituted soaps, are least stable, which readily dissociate with liberation of free fatty acid. The same soaps can be obtained, through precipitating them from an alkaline solution of sodium soap by an aqueous solution of aluminium sulphate. It is, however, difficult to obtain soaps stable in the composition and properties by these methods: the soaps produced are a mixture of mono-, di-, and trisubstituted soaps, whose ratio depends on many factors of their production technology.

Most suitable for grease production are disubstituted aluminium soaps, which are typical polymeric compounds [2, 12, 22-24]

Such soaps offer a high thickening effect and are therefore a valuable stock for production of complex aluminium greases.

Disubstituted complex aluminium soaps used in production of greases, is obtained by interaction of aluminium isopropoxide with carboxylic acids, such as stearic and benzoic acid, in an anhydrous medium, *i.e.*, in oil which serves as the dispersion medium of greases. The ratio of the reacting components, production-process regulations, and necessary (presented below) sequence of the reaction of complex soap formation should be strictly observed here;

(*a*) partial hydrolysis of aluminium isopropoxide (AIP):

$$Al(OC_3H_7)_3 + H_2O + \xrightarrow[oil]{90-95°C} AlOH(OC_3H_7)_2 + C_3H_7OH;$$

(*b*) production of a mono substituted soap of a high molecular weight fatty acid, such as stearic (R1COOH):

$$AlOH(OC_3H_7)_2 + R_1COOH \xrightarrow[oil]{120-125°C} Al\begin{matrix}OH\\-OC_3H_7\\OOCR_1\end{matrix} + C_3H_7OH; \text{ and}$$

(*c*) production of the basic disubstituted (complex) soap of stearic (R1COOH) and benzoic (R$_2$COOH) acids:

$$Al\begin{matrix}OH\\-OC_3H_7\\OOCR\end{matrix} + R_2COOH \xrightarrow[oil]{120-125°C} Al\begin{matrix}OH\\-OOCR_2\\OOCR_1\end{matrix} + C_3H_7OH$$

(complex aluminium soap)

A preferable molar ratio of the components forming the cAl-soap is:

$$AIP : H_2O : R_1COOH : R_2COOH = 1 : 1 : 1 : 1.$$

Lead soaps are produced [2, 22] using a double exchange reaction between the Na-soap of fatty acids and lead acetate in an aqueous medium:

$$2RCOONa + Pb(CH_3COO)_2 \cdot 3H_2O \rightarrow (RCOO)_2Pb + 2CH_3COONa + 3H_2O.$$

The reaction is conducted at 50-60°C under vigorous stirring. Precipitated Pb-soap is isolated from the mother liquor, washed with hot (50-60°C) water till absence of acetate ions, which is checked by the reaction with potassium dichromate, then dried by heating to 100-120°C and after the water removal fused at 130-140°C; at this temperature it is poured into moulds for cooling. Finished soap form can be used for grease production.

Physico-chemical characteristics of soaps. The crystal lattice structure, melting point, polymorphism and stepwise melting, *i.e.*, phase transitions, solubility and swellability in polar and nonpolar solvents, critical concentrations of association and micelle formation in hydrocarbons, capability of gel formation in various dispersion media, in particular in petroleum oils, and thermal decomposition, polarity and other properties of soaps substantially affect the structure of the skeleton and characteristics of greases. These properties of soaps depend on the cationic and anionic parts of their molecules, production methods, presence of foreign impurities and water, and when they are produced in situ, also on the composition of the dispersion medium and other factors. It is to be pointed out that many of the above-listed properties for various soaps have been in most cases inadequately studied. The data presented in literature are often contradictory.

We will now briefly discuss those features and properties of soaps [8, 11, 12, 22-38, 46] which should be taken into consideration in production of greases because they affect the formation of the structure of the latter.

Depending on the anion and cation used to prepare a soap, as well as on the content of acid or basic salts, water, and other impurities in it, soaps can be hard or soft suitable for forming or not forming the grease structure. Solid fatty acids form hard, and liquid ones soft soaps. The hardness of soaps depends also on the cation: e.g., potassium soaps are softer than sodium ones. All soaps are hygroscopic. Soaps of alkali metals are most hygroscopic and those of polyvalent metals, least hygroscopic. The hygroscopicity of a soap is also determined by its anion. Soaps of unsaturated acids are more hygroscopic than soaps of saturated acids. Fully dewatered soaps are very difficult to obtain practically. When absorbing moisture, soaps swell, by increasing the dried volume. The swelling is accompanied by heat release, which in storage of dry powdered soaps leads sometimes to charring and even ignition. The water-absorbing capability of soaps is called gelation. This capability for stearates of various metals declines in the order Na > K > Li > Ca > Pb [8].

The solubility of soaps in water and various hydro carbons is also determined by their composition. The water solubility of low molecular weight acid soaps is higher than that of high molecular weight ones. The solubility of high molecular weight acid soaps increases in presence of low molecular weight ones. The solubility of unsaturated acid soaps is more than that of saturated acid soaps. The solubility of soaps depends also on their cation, the best water solubility being shown by sodium soaps. The solubility of soaps increases with temperature. Aqueous solutions of soaps are electrically conducting. A certain electrical conductivity is exhibited by hydrocarbon systems of soaps.

The X-ray structure analysis of soaps of various metals shows that they have essentially similar structures of the crystal lattice [27]. Cations (polar groups) are arranged in parallel planes spaced from one another at a double length of soap anions, multiplied by the sine of the angle of their inclination with respect to the planes. The bonding forces between polar groups of soaps greatly exceed the interactions between end methyl groups of the hydrocarbon radicals. Thus, for lithium stearate the former is equal to 125.4 kJ/mole (30 kcal/mole) and the latter 1.25 kJ/mole (0.3 kcal/mole) [39].

Soaps of fatty acids form orthorhombic and monoclinic crystals. When soaps are heated till formation of an isotropic melt by them, a stepwise destruction of their crystal lattice occurs, and at definite temperatures the soap goes over from one mesomorphous phase to another. These phenomena are called phase transitions of soaps. Every soap has its characteristic number and absolute values of phase transition temperatures. However, a conclusive interpretation of the phase transitions of soaps is absent so far.

Best studied in the phase transitions are sodium soaps. The author of [27] distinguishes for anhydrous sodium soaps the following phases, listed in the order of increasing temperatures of their stable existence: crystalline; supercrystalline; prewax; wax; superwax; pretranslucent; translucent and isotropic melt. However, a detailed structure of individual phases has been inadequately elucidated. The study of the stepwise melting of soaps and of the structure of individual phases is difficult and calls for using different techniques to determine individual phase transitions. The difficulty of estimating the transitions stems also from a slow progress of some of them not only during the cooling of melts, but also during the heating of systems. Thus, the sodium palmitate transition from the crystalline to the prewax phase lasts over 2 hours. Phase transitions of soaps are greatly affected by their moisture content, presence of surfactants and other factors. The heating and cooling curves for alkali and alkaline earth metal soaps are in most cases not identical, the number of phase transitions during heating being greater than during cooling. Table 3.18 presents the phase transition and melting temperatures of some soaps [5, 12, 25-38] that are of practical importance in grease production. The character and absolute values of temperatures of phase transitions of soaps (Table 3.18) are not identical, but the data provide a general picture of mesomorphous states of soap crystals under the action of temperature. Some phase transitions of soaps are of particular importance in formation of the structure of greases. Thus, the $Ca(St)_2$ phase transition at 123-127°C involves the greatest thermal effects, which amounts to about 70% of the total thermal effect of all transitions [27]. At this transition $Ca(St)_2$ acquires the capability to swell in many hydrocarbons and to stretch into thin long filaments. The region of temperatures of structurization, of anhydrous Ca-greases corresponds to phase transitions of $12\text{-}Ca(St)_2$ at 105-110°C and 125-127°C [46].

The soap behavior in various solvents, in our case in hydrocarbon oils, is essential. The state and properties of systems consisting of a soap and a solvent depend on the temperature, concentration of components, their nature, presence of surfactants in the systems and other factors [7, 24-46]. The solubility of soaps in oil increases with temperature; at a temperature definite for every soap the solubility becomes infinite with the formation of an isotropic melt.

The solubility of soaps in oils increases with increasing valency and atomic mass of the cation. The presence of surfactants in the system intensifies the swelling and increases the

Table 3.18. Temperature of phase transitions and melting of some soaps

Cation	Temperature of phase transitions, ºC			Melting temperature, ºC	
	Stearate	12-oxystearate	Source	Stearate	Source
Li+	113, 185, 224	179, 218	[26]	220-221	[5]
		165	[27]	220.5	[38]
	122, 190, 219, 229	185, 218	[28]		
	115, 192, 220	128, 173, 205			
		212	[31]		
	115, 200	209	[32]		
	114, 187, 225	175, 220	[33]		
		185, 207	[34]		
	106, 173, 188, 206		[38]		
Na+	70, 89, 117, 132 167, 205		[27]	255-272	[12]
	89, 117, 132, 167 205, 257, 288	118, 125, 172 206, 230	[28]	260	[38]
	89, 134, 144, 169 208, 238, 280		[38]		
K+	69, 149, 170, 185 210, 225, 238, 272	98, 115, 147 185, 241, 260	[28]	150-156	[12]
Ca2+	90-110; 120-135; 160-170		[25]	350 (with decomposition) 150	[35]
	65, 86, 125, 150, 195		[27]		[38]
	95, 127, 155, 179 189, 350	128, 152, 225, 295	[29]		
	107, 130, 152	130, 150	[32]		
	65, 86, 123, 150, 195	110, 127	[34]		
	113, 125, 150	150, 214	[35]		
		105, 127, 147 209	[36]		
	65, 86, 123, 150 195		[38]		
Ba2+	110, 150, 200, 235		[25]		
	165, 263, 272, 337 355	170, 305, 325 335	[29]	Decomposes	[5]
	58, 136		[37]		
	58, 138		[38]		
Pb2+	69, 101, 133	—	[37]	116	[5]
	69, 101, 135		[38]	116	[38]
Al3+	60, 95, 150 (disubstituted)	—	[30]	Mono- : 170; di- : 145; tri- : 115; 113-173 (not specified)	[12] [38]
	99, 119, 162 (not specified)		[38]		

solubility in oils. The swelling of a soap in an oil is accompanied by its dispersion, which at some temperatures is ahead of the swelling process.

An important property of soaps is their ability for micelle formation in hydrocarbon (petroleum) oils, *i.e.* formation of crystallization centres [39-44]. A soap is characterized by a definite value of the critical micelle concentration (CMC), which depends on the cation and anion as well as on the dispersion medium properties. Thus, the CMC values for LiSt, NaSt and $Pb(St)_2$ in petroleum oil AU is 0.1; 0.2; and 0.3 mass % respectively [39]. The formation of micelles and aggregates of surfactants (soaps) in hydrocarbon media results from interaction of polar groups of molecules as well as from a dipole interaction through hydrogen bonds with emergence of specific coordination bonds, accompanied by increase of the entropy and decrease of the free energy of the system [41].

The swellability, solubility, polymorphism, energy of bond between polar and methyl groups of molecules, oleophilic-hydrophobic balance, CMC, energy of bonding of molecules into micelles, aggregation of micelles, adsorptivity and other characteristics of soaps substantially affect the structurization processes. This determines the selection of the optimum technology and influences the formation of grease properties. However, the soap properties widely used in the production of greases (particularly of soaps obtained in situ) have been inadequately studied.

3.2. SOLID HYDROCARBONS

Solid hydrocarbons—paraffins, petrolatums, ceresins and bitumens, contained in petroleum, as well as natural waxes, peat and lignite (montan) ozokerites, products of their processing, etc., whose properties are described in [47-62]—are used as thickeners in production of hydrocarbon greases serviceable up to 60-70°C and offering good protective properties.

The composition and properties of solid petroleum hydrocarbons depend on the final boiling point of the petroleum (oil) fraction, low-boiling oil fractions contain primarily solid paraffin hydrocarbons of a normal structure. As the boiling limits of fractions are increased, the content of n-paraffins in them decreases, while the share of isoparaffinic and cyclic, especially naphthenic hydrocarbons with long side chains increases. Naphthenic hydrocarbons with side chains, mainly of an isostructure, concentrate primarily in residual products. The latter also contain aromatic hydrocarbons with long alkyl chains of a normal and an isostructure and in an insignificant amount, high molecular weight isoparaffinic hydrocarbons. The methods of producing solid hydrocarbons from petroleum fractions rely mainly on the ability of these hydrocarbons to crystallize at low temperatures.

The removal of solid hydrocarbons from petroleum fractions is called deparaffination of petroleum products. It is effected by various methods: crystallization of solid hydrocarbons during cooling of the starting stock; crystallization of solid hydrocarbons during cooling of the solution of the starting stock in selective solvents; interaction with carbamide, complexation;

catalytic conversion of solid hydrocarbons into low-congealing products; biological action; adsorptive separation and other methods. Methods of deparaffination by selective solvents are most widely used. The methods of deparaffination of petroleum products and other methods for obtaining solid hydrocarbons as well as their properties are described in detail in [47-54]. The general scheme of producing oils from the petroleum stock, followed by isolation of solid hydrocarbons—paraffin, petrolatum and ceresin—is shown in Fig. 2.1.

3.2.1. Paraffins

Production of paraffins [47-53] is based on deoiling of batches formed in deparaffination of distillate oils. This yields medium-melting (melting point, 45-50°C) and high-melting (50-65°C) solid paraffins. They are also separated out by deparaffination of paraffinic distillates, specially obtained for this purpose from low-sulfur paraffinic petroleums. Deparaffination of diesel fuels and other petroleum fractions yields soft (melting point, 28-45°C) and liquid (melting point below 27°C) paraffins.

Solid paraffins are of practical interest for production of greases. These are crystalline products containing 20 to 40 carbon atoms in the n-alkane chain. Apart from n-alkanes (75-98 %) solid paraffins contain isoalkanes (2-25%) and a very insignificant amount of naphthenic and aromatic hydrocarbons with long paraffinic side chains.

A wide range of solid paraffins is commercially produced to GOST 23683-79: three grades of highly refined food paraffins (P-1, P-2, and P-3), differing in the melting point and oil content (the melting point for P-1 should be not below 54°C; for P-2, not below 52°C; and for P-3, not below 50°C; the oil content, not over 0.45, 0.9 and 2% respectively); five grades of highly refined paraffins (V_1, V_2, V_3, V_4, and V_5), differing mainly in the melting point and needle penetration at 25°C (the melting point for the B_1 grade is of 50-52°C; for B_2, 52-54°C; for B_3, 54-56°C; for B_4, 56-58°C; and for B_5, 58-62°C; the needle penetration under a load of 100 g should not exceed 18, 16, 14, 13, and 12×10^{-4} m respectively); commercial refined paraffins of grade T for various purposes and grade S for synthesis of SFA; and two grades of nonrefined paraffins (N_s for production of batches and N_v—nonrefined high-melting one for production of α-olefins and other purposes). Commercial refined paraffin T, nonrefined high-melting paraffin N_v, highly refined paraffin V_5 and food-grade paraffin P-2 are of the greatest interest for grease production. Characteristics of paraffins are presented in Table 3.19.

Commercial (T) and high-melting (N_v) paraffins are used to produce general-purpose greases, while highly purified (V_5) and food-grade (P-2) ones, for special instrument and other greases (jellies) as well as for greases used in the foodstuff production. However, paraffin has not found wide application in production of greases, which stems from its relatively low melting point, high brittleness at low temperatures, and relative scarcity. Paraffin is sometimes used in combination with other thickeners in grease production.

Table 3.19. Characteristics of solid paraffins used as dispersed phase of greases (GOST 23683-79)

Characteristics	Paraffin grade			
	T	N_v	B_5	P-2
Appearance	White crystalline mass	Crystalline mass of light-yellow to light-brown color	White crystalline mass	White crystalline mass
Melting point, °C (not below)	50	57	58-62	52
Oil content, % (not over)	2.3	2.3	0.5	0.9
Needle penetration depth at load of 10,000 Pa, 10^{-4} m (not over)	—	20	12	—
Flash point (in closed crucible), °C (not below)	—	180	—	—
Content, % (not over):				
sulfur	—	0.15	absent	—
water	absent	traces	absent	—
mechanical impurities	absent	0.01	absent	—
phenol	—	—	absent	—
sulfates and chlorides	—	—	absent	—
acids and alkalis	—	—	absent	—
benzopyrene	—	—	absent	—
Color by glass No.1, mm (not below)	—	—	—	250
Color stability, days (not below)	—	—	—	7

3.2.2. Petrolatums

Petrolatums are produced, during dewaxing of residual oils [47, 49, 53, 57]. They are mixtures (solutions) of high-melting paraffin and isoparaffins, naphthenes and aromatics with long side chains (predominantly isostructural ones) and other solid hydrocarbons in residual oils. The ratio of the above-stated hydrocarbon groups in petrolatums varies with the chemical nature of the stock processed. The solid part of a petrolatum usually contains 80-90% paraffin-naphthene hydrocarbons, 3-10% aromatic hydrocarbons and 2-4% asphaltic-resinous compounds; the share of residual oils in petrolatum amounts to 20-25%. Petrolatum is made as a commercial product according to GOST 38-01117-76. Depending on the method of refining (deasphalting, dearomatization) of the starting stock in production of residual oils and on the composition of petroleum from which the tar has been obtained (Low-sulfur or sulfurous), three petrolatum grades are produced: PK (sulfuric acid refined); PS (selective-refined); and

PSs (selective-refined, from sulfurous petroleums). Production of the PK grade petrolatum is very limited. Characteristics of petrolatums are presented in Table 3.20.

Table 3.20. Characteristics of petrolatums (GOST 38-01117-76)

Characteristics	Petrolatum grade		
	PK	PS	PSs
Color	Light-brown	Light-brown	Light-brown
Drop point, °C (not below)	55	55	55
Test for corrosive action	Withstands	Withstands	Withstands
Acid number, mg KOH/g (not over)	0.05	0.1	0.1
Content, % (not over):			
water-soluble acids and alkalis	absence	absence	absence
mechanical impurities	0.04	0.03	0.04
water	0.9	traces	traces
sulfur	-	-	0.06
phenol	-	absence	absence
Flash point in open crucible, °C (not below)	255	240	230
Kinematic viscosity, $\times 10^{-6}$ m^2/s (not below)	-	-	16
Maximum permissible concentration of hydrocarbon vapors in working zone air, mg/m^3 (not over)	300	300	300

Table 3.21. Group composition of ceresins

Ceresin	Paraffin-naphthene hydrocarbons		Aromatic hydrocarbons	Resinous substances
	complexing with urea	not complexing with urea		
Volgograd-field petroleum ceresin, grade 80	26.1	67.1	4.0	2.8
Grade 80, obtained from Borislav-field ozokerite	39.7	49.6	8.2	2.5

Petrolatum is used for grease production individually or in combination with other thickeners and also, after its oxidation, as adhesive and protective additives. Oxidized petrolatum is occasionally used as a saponifiable component. Petrolatum is employed as a thickener for production of protective hydrocarbon greases such as commercial petroleum jelly, rope and other greases. It also serves as the starting stock to produce petroleum ceresin.

3.2.3. Ceresins and Petroleum Waxes

Deoiling of petrolatum yields unpurified petroleum ceresin, which is then subjected either to a sulfuric acid or to a selective and an adsorption refining to obtain corresponding grades of refined ceresins [47, 52, 53, 57]. Other ceresin production sources are wax accumulations in the petroleum production (paraffin plug) and natural ozokerites. Ceresins can also be obtained by the synthesis of carbon oxides and hydrogen; this yields high-melting synthetic ceresins, which are mixtures of solid (paraffin) hydrocarbons of predominantly normal structure. They have found no practical use in grease production because of their tendency to gel on cooling.

Petroleum ceresins and ceresins resulting from the processing of paraffin plugs and ozokerites are crystalline products whose colour varies from yellow to brown. Their composition includes primarily naphthenic and aromatic hydrocarbons with long paraffinic side chains, mainly isostructural as well as high-molecular isoalkanes and insignificant amounts of normal paraffins. The composition of a petroleum ceresin somewhat differs from that of a ceresin produced from ozokerite (Table 3.21).

The melting point of ceresins depends on the composition of the processed stock and on the refining method and varies from 57 to 85°C; their molecular weight varies from 500 to 750. Ceresins differ from paraffins of the same molecular weight by higher melting point, viscosity, density, elasticity and reactivity as well as by higher thickening efficiency, tackiness, adhesion, and mechanical strength. This encourages an extensive use of ceresins as thickeners to produce hydrocarbon greases. They are also used in mixtures with other thickeners.

A broad range of ceresins are commercially produced (Table 3.22). Note that petrolatums and paraffin plugs are at present a basic stock for production of ceresins, while the use of ozokerites for their production is very limited.

3.2.4. Petroleum Bitumens

Petroleum bitumens are employed, to produce greases for limited applications [52, 54, 55]. A practical application has mainly been found of building bitumens, produced according to GOST 6617-76 in three grades: EN 50/50, EN 70/30, and BN 90/10 (Table 3.23). Building petroleum bitumens are produced by oxidation of residual products of a straight-run petroleum distillation and of their mixtures with asphalts and extracts of the oil production. In the production of building bitumens it is allowed to mix above-mentioned oxidized and unoxidized petroleum products.

3.2.5. Natural Waxes

The concept "waxes" here does not have the chemical meaning, according to which the chemistry of fats defines waxes as mixtures of esters of high-molecular fatty acids and aliphatic mono- and dibasic alcohols regardless of their state of aggregation at room temperature, but has the physical meaning. Classed with waxes are here such complex organic-origin substances which in respect of physical properties and in the scope of their application are very close to bees-wax, the origin of this term.

From the standpoint of chemical structure, such substance as lanolin, wool wax, are typical representative of "chemical waxes", while ozokerite does not fall into this class of compounds, but even an expert would call wax the ozokerite than the salve like lanolin.

A wide range of natural waxes are used in domestic and foreign practice as thickeners or additives improving some or other grease properties. A practical use in production of general-purpose greases from natural waxes has been found in ozokerite, lignite (mineral) and peat waxes.

Table 3.22. Characteristics of ceresins

Characteristics	Unrefined petroleum ceresin (TU 38.101507-79)		Refined ceresin (GOST 2488-79)				High-melting synthetic ceresin (GOST 7658-79)	
	75/80	80/85	80	75	67	57	Caracitor-grade	100
Melting point, °C (not below)	75-80	80-85	80	75	67	57	100	100
Needle penetration depth at 25°C and load of 10,000 Pa, 10^{-4} m (not over)	14	12	16	18	30	30	10	10
Content*: di- mechanical impurities, % (not over)	0.1	0.1	0.02	0.05	0.05	0.05	Absence	0.08
dii- sulfur, % (not over)	0.3	0.3	—	—	—	—	—	—
Acid number, mg KOH/g (not over)	0.25	0.15	0.10	0.15	0.15	0.15	0.05	0.18
ASH content, % (not over)	0.15	0.10	0.02	0.02	0.02	0.02	0.02	0.02
Electrical volume resistivity at 100°C, ohm•cm (not below)	—	—	$1 \cdot 10^{12}$	—	—	—	$1 \cdot 10^{14}$	—
Flash point, °C (not below)	290	290	—	—	—	—	—	—

*Content of water, water-soluble acids and alkalis, and phenol: absense.

Ozokerite [56-58] belongs to bituminous minerals. Ozokerite ore deposits primarily occur adjoining high-paraffin petroleum fields.

The most valuable part of ozokerites are solid hydrocarbons, whose content in ozokerites of various deposits, including various petroleum samples, varies from 15 to 90%. The less valuable part consists of liquid and soft hydrocarbons (8-80%) with melting points below 50°C and resinous compounds (6-30%) [57].

The properties and composition of ozokerites differ with their deposits. Thus, solid hydrocarbons with a melting point of 77-79°C were separated from vein ozokerites of the Shor-Su deposit and with a melting point of 71-87°C from bed ozokerites of the same deposit. Ozokerite from the Borislav deposit contains solid hydrocarbons with a melting point of 85-90°C, their content in raw ozokerite being 65% on an average.

Table 3.23. Characteristics of building petroleum bitumens (GOST 6617-76)

Characteristics	Bitumen grades		
	BN 50/50	BN 70/30	BN 90/10
Needle penetration depth at 25°C, 10^{-4} m	41-60	21-40	5-20
Softening point by ring and ball, °C (not below)	50	70	90
Stretchability at 25°C, cm (not below)	40	3	1
Solubility in benzene or chloroform, % (not below)	99	99	99
Flash point, °C (not below)	220	230	240
Content of water-soluble compounds, % (not over)	0.3	0.3	0.3
Content of water	traces	traces	traces

Ozokerite is at present commercially produced from Borislav's ores. According to STP 39.8637.20-79 it should meet the following standards: drop point not below 55°C; penetration at 25°C, $(30-100) \times 10^{-4}$ m; content of mechanical impurities, not over 1%, water absent; water-soluble acids and alkalis: absent. Such ozokerite is used in production of rope greases.

Lignite wax [58, 60], sometimes called mineral or montan wax, is produced from lignites by extraction with various solvents. Its properties are determined by the composition of lignites, dissolving power of solvents used and extraction process conditions. Lignite wax is produced on a commercial scale from ground lignite by treating it at 80°C with benzene or gasoline. Extracted lignite wax after removal of the solvent is a dark-coloured solid brittle product having a high (up to 90°C) melting point, low electrical conductivity, good mechanical strength, high chemical stability and thickening efficiency; its saponification number is 70-95 mg KOH/g and iodine number 8-15 mg of iodine per 100 g of the product. Raw lignite wax is produced by the method of gasoline extraction from lignites. According to TU 39-01-232-76, it should meet the following standards: drop point, not below 82°C; content of substances insoluble in gasoline, not over 0.35%, ash content, not over 0.1%, gum content at 20°C not over 20%; fusibility with paraffin (at 1:3 ratio) complete; no separation of solvent on melting.

Two parts are conventially distinguished in lignite wax, a wax part (wax proper) and gum, the ratio between which depends on the stock composition and wax production technology. The wax part contains mainly esters of saturated fatty acids C_{20}-C_{34} and of monobasic alcohols C_{24}-C_{30} (50-60%) as well as free fatty acids C_{17}-C_{34} (15-25%), alcohols C_{16}-C_{32} (15-20%), lactones and other organic compounds (3-8%) [58, 60]. Gum of lignite wax contains highmolecular weight acids and alcohols, triterpenic ketones and lactones and complex aromatic hydrocarbons with various functional groups.

The rich chemical composition of lignite waxes allows their use not only as a hydrocarbon thickener in production of, *e.g.*, rope greases, but also as an additive upgrading the protective, antiwear and antiseizure properties of greases. After a special treatment (deresining, etherification, refining, etc.), lignite wax (montanic acids) is widely used in Germany as a saponifiable component in production of high quality greases.

Peat wax [58, 61] is produced from bituminous peat by treating it, after an appropriate preparation (drying, grinding) with gasoline at 80-85°C. After the extraction and distillation of gasoline, the raw peat wax produced in the erst while USSR is a solid dark-brown product consisting 99% of high molecular weight organic compounds. According to TU 6-01-973-73 peat wax has the following characteristics: drop point, 70-80°C; content of resinous compounds soluble in acetone, not over 40%; of water, not over 0.15% and of mechanical impurities, not over 1%, fusibility with paraffin (1:10), complete; acid number, 30-60 mg KOH/g; saponification number, 100-160 mg KOH/g; and iodine number, 15-30 g of iodine per 100 g of wax.

The peat wax composition varies depending on the starting stock composition, extractant and extraction conditions. The wax part contains esters of fatty acids C_4-C_{32} and of alcohols C_{20}-C_{30}, the same acids and alcohols in a free state, hydrocarbons C_{23}-C_{33} (paraffins, ceresins) and other complex compounds. The resinous part contains saturated and unsaturated straight-chain and branched chain aliphatic and terpenic hydrocarbons, alcohols and acids C_{16}-C_{26} as well as oxyacids, stearin and other inadequately studied substances. Esters of acids C_{24} and C_{26} prevail among esters of peat wax. Free acids are primarily C_{16}, C_{18}, C_{19}, C_{20}, C_{22} C_{24}, and C_{26} ones. Along with saturated straight-chain and branched chain acids, oleic acid has also been found in the peat wax composition. The unsaponifiable part of peat wax, apart from high molecular weight normal-structure and isostructural paraffin hydrocarbons and cyclic hydrocarbons, contains also free alcohols, mainly arachidic ($C_{20}H_{42}O$), lignoceryl ($C_{24}H_{50}O$), neoceryl ($C_{25}H_{52}O$), and insignificant amounts of other ones. The molecular weight of the alcohols varies within 340-370.

Treatment of raw ground peat wax with gasoline and cooling of the solution to minus 5°C yields deresined wax with a resin content not over 10%. The refining (treatment by selective solvents, by oxidants, distillation, etc.) of deresined peat wax, its etherification and other processing methods yield various grades of refined waxes, containing over 90% wax acids.

Raw peat wax is used as the thickening component of rope greases. It also improves the adhesive, antiseizure, antiwear, and anticorrosion properties of these greases. Deresined peat wax and its refined grades can, apart from the above-mentioned functions, serve as a saponifiable component of greases, but data on studies of peat wax as a saponifiable component of greases are practically absent.

3.3. INORGANIC THICKENERS

Solid inorganic materials having a high dispersity, hydrophobicity, ability to form the primary skeleton of grease and to restore rapidly the bonds between particles after their breakdown on deformation can be used as thickeners in the grease production. Such a use has been found by highly dispersed modified silica, clay minerals featuring a high exchange capacity, lyophilic graphite, asbestos, and other highly dispersed minerals having the above properties.

3.3.1. Highly Dispersed Silica

Greases produced by thickening of lubricating liquids by highly dispersed silica (SiO2) are called silica gel greases. Highly dispersed silica can be obtained by several methods, most widely used of which are the pyrogenic and precipitation methods.

The pyrogenic method of production of highly disperse SiO_2 (aerosil) is based on a high-temperature (1000-1100°C) hydrolysis of silicon tetrachloride by steam forming on combustion of hydrogen in presence of oxygen [63, 64], which yields highly dispersed SiO_2 by the reaction

$$SiCl_4 + 2H_2O \rightarrow SiO_2 + 4HCl$$

Reaction products after cooling to 300°C pass through a coagulation zone where minutest SiO_2 particles, colliding with one another, form flakes of 1-2 nm. Chloride adsorbed by the SiO_2 surface is removed by treating SiO_2 in absence of air.

Aerosil is a very pure nonporous SiO_2 featuring a high dispersity, uniformity of particle sizes and a high specific surface, from 150 to 450 m²/g [64]. The particle sizes, which are from 5 to 40 nm (mainly 10-15 nm), depend on the flame temperature, $SiCl_4$ concentration, time of residence of SiO_2 particles in the reaction zone and other process factors. This product contains about 99.8% SiO_2, the balance consisting of oxides of aluminium, titanium, boron, iron and of other elements. Aerosil is commercially produced at the Kalush "Khlorvinil" PA to GOST 14922-77 in three grades, A-175, A-300, and A-380, whose main characteristics are presented in Table 3.24.

Table 3.24. Characteristics of highly dispersed silicon dioxide, aerosil (GOST 14922-77)

Characteristics	Aerosil grade		
	A-175	A-300	A-380
Appearance: uncompacted compacted	Light-bluish-white loose powder White mass in form of small loose lumps		
Content of silica, referred to calcined product, mass % (not below)	99.85	99.85	99.80
Total content of sesquioxides, mass % (not over)	0.083	0.083	0.093
PH of suspension	3.6-4.3	3.6-4.3	3.6-4.3
Content of hydrochloric acid, mass % (not over)	0.025	0.025	0.025
Moisture content, mass % (not over)	2	3	4
Mass loss in calcination, mass % (not over)	2	2	2
Bulk density, g/l: uncompacted compacted	40-60 120-140	40-60 110-140	40-60 -
Specific surface area by BET method, m²/g	175±25	300±30	380±45
Content of coarse particles (grit), mass % (not over)	0.04	0.04	0.05

Production of highly disperse SiO_2 by the precipitation method is based on the reaction of an aqueous solution of sodium silicate with sulfuric acid or ammonium sulfate:

$$Na_2SiO_3 + H_2SO_4 \rightarrow SiO_2 + Na_2SO_4 + H_2O,$$
$$Na_2SiO_3 + (NH_4)_2SO_4 \rightarrow SiO_2 + Na_2SO_4 + 2NH_3 + H_2O.$$

Production of highly dispersed SiO_2 by the precipitation method is usually combined with the process of its surface modification by organic compounds, such as butyl alcohol [65-67]. The production technology is described in [65, 66].

The surface of aerosil and of highly dispersed SiO_2 produced by the precipitation method is covered by hydroxyl groups, which imparts a hydrophilic character to the surface. The structure of the hydrated layer is fairly complex [68-70]. Main structural elements of the surface of highly dispersed SiO_2, actively participating in adsorption and chemical interactions are hydroxyl functional groups and silicon atoms carrying the hydroxyl groups. The latter are arranged not at every silicon atom of the SiO_2 surface, but at every alternate atom. Considerable distances between hydroxyl groups impede formation of hydrogen bonds. Along with hydroxyl groups, also coordination-bonded water, whose molecules are protonated and can interact with other water molecules through formation of hydrogen bonds is attached to silicon atoms. According to [70], hydro-philic aerosil has the following formula:

Such aerosil and highly disperse SiO_2 produced by the precipitation method are unsuitable to produce Si-greases since they are not moisture-resistant, but are hydrophilic. To improve the moisture resistance of such greases, highly disperse SiO_2 is hydrophobized by subjecting it to a chemical modification by surfactants.

Surfactants of various classes are recommended as hydrophobizers for the surface of highly dispersed SiO_2: primary and secondary alcohols, aliphatic amines and quaternary ammonium compounds, ketones and ethers, polyamines, their salts and amides, organosilicon compounds, and compounds containing phosphorus and sulfur, isocyanates and epoxy compounds, polymers, and other substances [15,71-88]. However, the widest practical use in production of modified silicas, thickeners for silicagel greases, has been found by dimethyl-dichlorosilanes and n-butyl alcohol.

The aerosil surface modification degree, kinetics and mechanism of reactions proceeding on the surface, and structure of compounds formed by interaction of n-butanol or dimethyldichlorosilane with SiO_2 depend on the state of the hydrophilic aerosil surface and on many factors of its modification process. Thus, controlling the surface hydration degree (the molecular water content in the surface layer) by the temperature of the vacuum pretreatment of silica makes it possible to obtain methyl aerosil derivatives containing various amounts of dimethylsilyl groups in the modified layer: less than the concentration of structural hydroxides; their full substitution with modifying groups; or even exceeding the concentration of hydroxyl groups on the initial aerosil surface [68-70].

Properties of silica gel greases are substantially dependent on the SiO_2 surface properties, in particular on the concentration of surface hydroxyl groups and of modifying groups, pH of the alcoholic-aqueous extract of aerosil, amount of physically and coordination bonded water, and other factors. SiO_2 modified by dimethyl-dichlorosilane should be used to produce thermally stable (up to 180-200°C when based on petroleum oils such as MC-20 and up to 250-270°C when based on polyphenylmethylsiloxane oils) silica gel greases, and SiO_2 modified by hexamethylcyclotrisiloxane-type compounds, for greases with a higher oxidation stability. The content of dimethylsilyl groups on the surface of modified SiO_2 should be not less than 1.1 mmole/g SiO_2; the concentrations of physically adsorbed and coordination bonded water should not exceed 0.6 and 0.5 mmole/g SiO_2 respectively; and pH of its alcoholic-aqueous extract should be within 5.6-6.2. When butoxy aerosil is used (greases are serviceable up to 130-135°C), the content of butoxy groups on its surface should be not less than 0.44 mmole/g SiO_2; the concentrations of physically adsorbed and coordination bonded water should not exceed 0.68 and 0.30 mmole/g SiO_2 respectively; and pH of an alcoholic-aqueous extract of SiO_2 should be within 5.6-6.2.

Modified aerosils—methyl aerosil (grades AM-1-175 and AM-1-300) and butoxy aerosil (butosil, grades B-1 and B-2) are commercially produced at the Kalush "Khlorvinil" PA according

Table 3.25. Characteristics of aerosil modofied with dimethyldichlorosilane (TU 6-18-185-79)

Characteristics	*Aerosil grade*	
	AM-1-175	*AM-1-300*
Appearance	Free-flowing powder without foreign inclusions, from white to light-grey in color	Free-flowing powder without foreign inclusions, from white to light-grey in color
Hydrophobicity, % (not below)	99.0	99.0
pH of suspension (not below)	3.6	3.6

Bulk density, g/l:
  uncompacted
  compacted
	
	25-50
	50-100
		
		25-50
		50-100

Content, mmole/g, of:
  hydroxyl groups (not over)
  dimethylsilyl groups (not below)
	
	0.05
	0.4
		
		0.05
		0.6

to TU 6-18-185-79 and TU 6-18-159-78; their characteristics are presented in Tables 3.25 and 3.26. Properties of highly dispersed silica (thickener BS) obtained by modification of precipitated SiO_2, which is commercially produced at the Moscow petroleum-oil factory, are presented in Table 3.27.

Table 3.26. Characteristics of highly dispersive silica (STP 3-1-75)

Characteristics	Aerosil grade	
	B-1	B-2
Appearance	Free-flowing powder without foreign inclusions, from white to light-grey in color	Free-flowing powder without foreign inclusions, from white to light-grey in color
Hydrophobicity, % (not below)	99.0	99.0
pH of suspension (not below)	4.8	4.8
Bulk density, uncompacted, g/l	20-45	20-45
Time of full destruction of hydrophobic coating in boiling, h (not below)	8.0	8.0
Content, mmole/g, of:		
hydroxyl groups (not over)	0.05	0.05
butoxy groups (not below)	0.4	0.5
Mass loss in calcination, % (not below)	2.5	3.5

Table 3.27. Characteristics of highly dispersive silica (STP 3-1-75)

Characteristics	Rating
Appearance	Irregular-shaped granules
Water stability, mass % (not below)	80
Mass loss in calcination, mass %	11-22
Bulk density, g/l	1-22
Ratio of mass loss in calcination to bulk density (not below)	50

Greases can also be produced with the use of aerosil modified by organic compounds containing nitrogen, sulfur and phosphorus, which improves the antioxidation, antiwear, antiseizure and other grease properties [77,84]. In operation of these greases, however, such a surface coating of aerosil breaks down progressively and greases lose their moisture resistance [7].

3.3.2. Clay Minerals

Organosubstituted bentonite clays are used in grease production [2, 7, 89-91]. Apart from them, also other clays are employed: attapulgite (palygorskite) [2,92], hectorite [89], vermiculite [93], etc. Greases produced by thickening of lubricating liquids by organosubstituted forms of clay minerals are called bentone greases. The use of clay minerals as the starting stock to produce grease thickeners is important both technically and economically since this provides for production of greases serviceable over a broad temperature range and at the same time obviates the need for natural fats and other scarce materials in their production.

Properties of starting clay minerals used to produce thickeners are essential not only in the modification of their surface, but also in the subsequent production of greases. Some characteristics of clay minerals are presented below.

Bentonite clays [94-97] are clay minerals with predominance of montmorillonite (50-100%) or beidellite, sometimes with admixtures of zeolite-group minerals. Montmorillonite belongs to the group of layered aluminosilicates with an expandable structural cell and a very imperfect structure. It consists of shapeless plastic particles with indistinct facets and great disorders in the crystal structure. Fine flakes of montmorillonite are of an imperfect morphology and strongly deformed. Basal planes of particles of mineral carry cations (Ca, Mg, Na, or K) which are readily substituted with organic and inorganic ions. This property of bentonites is utilized in production of grease thickeners based on them.

Palygorskite [94] belongs to the group of layered-ribbon sepiolitic-palygorskite minerals. Its main structural element consists of silicon-oxygen tetrahedrons arranged in the form of a double chain of the Si_4O_{11} composition. Palygorskite particles are needle-shaped. Needles aggregate into packs, whose porous space and zeolite channels determine the adsorption properties.

Kaolinite [95] is a layered mineral whose crystal structure features varying degrees of perfection and ordering in the relative orientation of basic networks in a layer and shift of individual layers with respect to one another.

Vermiculite [96, 98, 99] belongs to layered silicates. It consists of coarse lamellar particles. The microstructure exhibits aggregativity, aggregates being of an elongated shape. The surface of crystals is smooth and fairly uniform. The energy of interaction of exchange cations with elementary packs in vermiculite is higher than in montmorillonite and possibilities for expansion of a structural cell are correspondingly lower.

Some physico-chemical characteristics of clay minerals are presented in Table 3.28 [96, 100].

Table 3.28. Characteristics of clay minerals

Mineral, deposit	Structure type	Particle shape	Particle size, μm	Cation exange capacity, mg-equiv per 100 g of clay	Heat of wetting (by water), J/g	Specific surface, m²/g	
						N_2	H_2O
Montmorllonite: -Cherkassky -Askansky -Sarygyukhsky -Pyzhevsky -Oglanlinsky	Layered aluminosilicate with expandable structural cell;	Platelets with indis-tinct faceting	0.05-0.80 0.03-0.50 0.05-0.10 0.05-0.10 0.30-0.50	70.0 94.0 72.0 100.0 81.5	101.2 50.2 53.9 148.4 92.8	60 57 45 39 -	320 440 312 428 202
Palygorskite: -Kurtsevsky -Cherkassky	Layered-ribbon silicate with rigid crystal lattice	Acicular particles	0.2-0.5 (length) 0.005-0.1 (thickness)	24.5	156.3	7	302
Caolinite: -Glukhovetsky -Glukhovsky	Layered silicate with rigid lattice; 1:1 dioctahedral	Platelets with distinct faceting	1.00-2.00 0.05-0.20	1.5 25.5	5.4 40.1	9 70	12 94
Vermiculite -Kovdovsky	Layered silicate with expandable structural cell; 2:1 trioctahedral	Platelets	1.50-3.00	135.0	193.5	11	470

Cations adsorbed on the surface of clay minerals endow clays with the ability to form derivatives with organic substances. Such an interaction, *i.e.*, modification of the surface of clay minerals with organic compounds, results in organophilic and hydro-phobic complexes (organoclays) which produce colloidal systems featuring elasto-plasto-viscous and thixotropic properties in petroleum oils and synthetic lubricating liquids. This property of organoclays is utilized in producing thickeners for greases, formation of organoclay complexes depends first of all on the nature and composition of exchange ions of the clay surface and on the magnitude of their capacity. Sodium cations more readily enter into an exchange reaction with organic cations, and therefore in production of clay organo complexes the Ca, Mg, Li, and C cations are first substituted by the Na cation. Furthermore, secondary particles that are retained as polyvalent cations which are on the surface of montmorillonite get broken down in dilute suspensions of the Na-form [101]. Apart from exchange reactions, physical adsorption of the organic substance by the mineral surface also occurs. The surface coverage is a function of the number and length of adsorbed organic ions; these factors determine both the arrangement of radicals on the mineral surface and the spacing of individual elementary layers. Orientation of hydrocarbon radicals on the surface of particles depends also on the crystal structure of minerals

DISPERSED PHASE OF LUBRICATING GREASES

[100]. When the hydrocarbon radical of the modifier is removed and as the number of adsorbed ions increases, the chains progressively straighten with respect to the surface of bentonite until they become perpendicular to it. This state corresponds to the maximum spacing of adjacent layers [100, 102]. If organic radicals occupy over 50% of the surface of crystals of the mineral, then adjacent silica layers cannot approach each other closer than to two diameters of the n-alkyl chain [103, 104]. The higher the possibility of disconnection and separation of clay layers, the greater the thickening power of their organoderivatives in hydrocarbons.

On the whole, the thickening power of organoclay complexes in lubricating liquids (oils) and properties of bentone greases depend on the crystal structure of the clay material; size and specific surface of particles; chemical nature of the modifier; modification degree; screening capacity of the modifier (cation) and structure of the adsorption layer [7, 91, 104]. Used as clay surface modifiers are various compounds:

1. Nitrogen-containing compounds of the fatty series: amines, amides, diamides, amidozolines, primary, secondary, and quaternary ammonium salts, and other compounds with 12 to 20 carbon atoms in the chain. Most widely used are quaternary salts, featuring high solubility and displacing ability, such as octadecylamine hydro chloride, aniline, amidozoline, dimethyl dialkyl ammonium chloride (DMDAAC), dimethyl dioctadecyl ammonium chloride, dimethyl-octadecyl benzyl ammonium chloride [100-110].

2. Polymers and poly condensation, products, such as polyvinyl acetate, polystyrene, polyepoxy resins [111], products of condensation of epichlorohydrate with stearic acid amide and ammonia or tertiary amine, containing one alkyl and two polyoxyethylene groups [112] and asphaltenes [113].

3. High molecular weight compounds: phosphonium, arsonium, stibium, sulfonium, etc. [114].

The practical application of modification of clays for production of grease thickeners has been found by compounds listed in Table 3.29 [100].

Table 3.29. Characteristics of modifiers

Modifier	Chemical formula	Molecular weight (average)	Amount of active substance, mass%
Dimethylalkylbenzylammonium chloride (ABDM)	$\left[\begin{array}{c} CH_3 \\ \\ N^+ \\ CH_3 \quad C_nH_{2n+1} \end{array} \text{Ph} \right]^+ Cl^-$, $n = 16$ to 20	423	70
Dimethyldialkylammonium chloride (DMDAAC)	$\left[\begin{array}{c} CH_3 \quad R \\ N^+ \\ CH_3 \quad R \end{array}\right]^+ Cl^-$, $R = C_{14}$ to C_{18}	550	75

Aliphatic amine acetate (Armac NT)	$[R-NH_3]^+ CH_3COO^-$ $R = C_{16}$ to C_{18}	332	90
N-alkylpropylenediamine acetate (Duomac NT)	$[R-NH_2-(CH_2)_3-$ $-NH_2]^+ (CH_3COO)_2^-$	480	90
Octadecylamine acetate (ODA)	$C_{18}H_{37}NH_2{}^+CH_3COO^-$	328	—

Modification of the surface of clay minerals by organic compounds is effected by the following scheme:

$$Na^+-bentonite^- + \begin{bmatrix} R_2 \\ | \\ R_1-N-R_1 \\ | \\ R_3 \end{bmatrix}^+ -Cl^- \leftrightarrows \begin{bmatrix} R_2 \\ | \\ R_1-N-R_4 \\ | \\ R_3 \end{bmatrix}^+ -bentonite'^- + NaCl.$$

The process of modification can be accomplished by the following methods: cation exchange reaction in an aqueous medium; ionic exchange in an organic solvent; an exchange in vapour phase at elevated pressures and high temperatures.

To produce bentone greases, it is expedient to use ABDM-modified mixture of clay minerals, consisting of Sarygyukh montmoril-lonite (A) and palygorskite (B), with a surface coverage not less than 100 mg-equiv per 100 g of clay and a particle size of clay organoderivatives not over 0.250 mm; the A:B mass ratio is preferably within (6-7) : (3-4) [7, 15].

It should be noted that bentonite thickeners do not melt at the temperatures of grease applications (right up to 400°C), but the upper temperature limit of application of bentone greases is not over 150°C [1]. Bentonite thickeners in a mixture with soaps are often used to produce high-efficiency greases.

The technology of production of organobentonites by the method of cation exchange in an aqueous medium consists of the following main stages [2, 36, 107, 108]: preparation of a highly dispersed suspension of clay (bentonite); preparation of the modifier solution; bentonite surface modification; washing; filtration and drying of organobentonite. Organobentonites for use as the dispersed phase of greases are sometimes produced in grease plants. The flow diagram of this process is shown in Fig. 9.2.

Preparation of aqueous suspension of clay. To disperse the bentonite, it is treated with water, which penetrates into the interlayer space of bentonite, the latter swells and separation of particles results in a homogeneous fine suspension of monomineral. To do this, demineralized water from measuring tank T-1 is charged into vessel V-1 and heated to 40-50°C, after which the clay is charged into the vessel in small amounts under stirring. The water-to-clay ratio is about 20 : 1. When the clay charging has been completed, the stirring is stopped and clay swells in 30-35 min. Next the contents of vessel V-1 are heated to 65-70°C and stirred till the formation of a homogeneous suspension, which is then delivered by pump P-1 through screen filter F-1 into vessel V-2, from which the filtrate flows by gravity into separator S-1 for removal of impurities that have passed through the screen filter. The separation (centrifurging) results in two fractions: the desired fraction and the substandard one which needs reprocessing. The desired fraction (suspension) is collected in tank T-2, from which it is fed by pump P-3 into reactor R-1. The substandard fraction (suspension) flows down into tank T-3 and is then fed by pump P-2 through filter F-1 into vessel V-2.

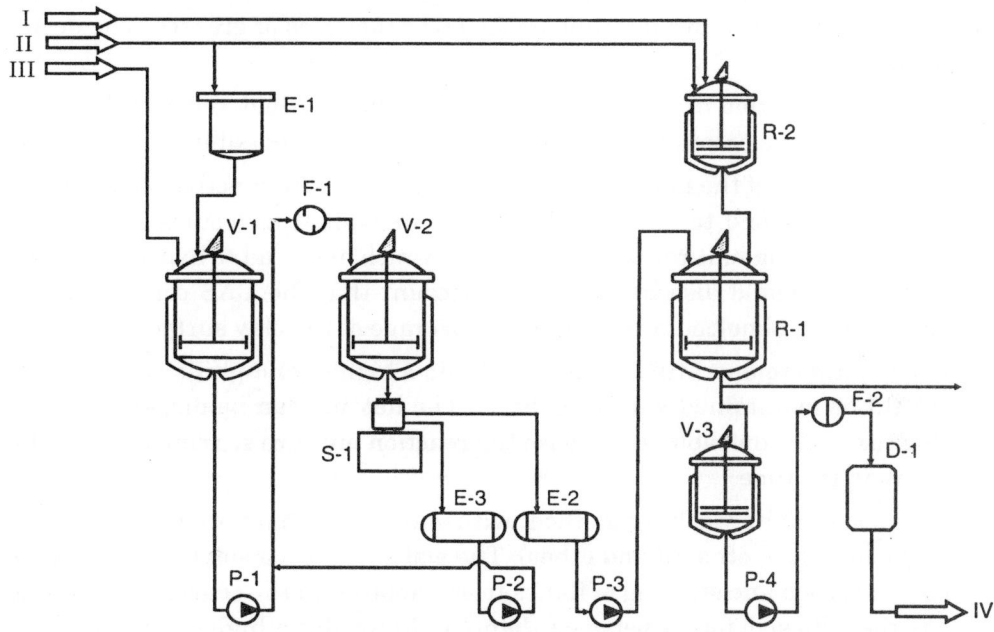

Fig. 3.2. Flow diagram of preparation of organobentonites:
T-1—measuring tank; E-2, 3—intermediate horizontal tanks;
V-1, 2 and R-1—vessels (reactors) with turbine-type stirrer;
V-3 and R-2 vessels (reactors) with paddle-type stirrer;
S-1—separator; D-1—drying chamber; F-1, 2—filters;
P-1, 2, 3, 4—pumps; I—modifier; II—demineralized water; UI—clay;
IV—finished product.

Preparation of modifier solution. Demineralized water of 1/3 of the calculated amount is filled into reactor R-2 fitted with a stirrer and a jacket for the heat carrier and then the calculated amount of the modifier is metered into the reactor. The solution is heated to 60-65°C under stirring (for about 30 min) while the remaining amount of water is added slowly. The water-to-modifier ratio (by mass) is of about 45 : 1.

Bentonite modification. The calculated amount of the modifier solution from reactor R-2 is charged slowly into reactor R-1 containing the standard-quality aqueous suspension of bentonite. The clay-to-modifier ratio in terms of the desired products is about 3:1. The content of reactor R-1 is heated to 50-55°C under stirring; an ion-exchange reaction resulting in formation of a modified bentonite (organobentonite) proceeds at this temperature. The reactor contents are then allowed to settle, with the result that organobentonite accumulates in the top layer, while the bottom layer of mother liquor is drained into the waste disposal system. The top organobentonite layer is collected into vessel V-3.

Organobentonite drying. Wet organobentonite (suspension) from vessel V-3 is delivered by pump P-4 through filters F-2 into drying chamber D-1 where temperatures of 150-160°C at the inlet and of 80-90°C at the outlet are maintained. After moisture removal, organobentonite is delivered for packing.

Specific features of the clay modification technology substantially affect properties of organobentonites, their structure forming ability and properties of greases. The deciding factors are as follows:

1. Clay concentration in water, which should ensure a rapid and complete proceeding of the exchange reaction and a high degree of clay dispersibility.

2. Concentration of the modifier, *i.e.* the solution of a cation-active substance (CAS); it should correspond to such a value where individual molecules are adsorbed in a considerable amount rather than micelles which may lead to hydrophilization of the surface. It should also be taken into account that the CAS concentration should provide for a bimolecular cation layer coverage of the clay surface.

3. Modification temperature, which, to obtain a highorgano-philicity organobentonite, should be maintained within 50-55°C. The heating can be dispensed with, but in this case it is advisable to increase the reaction mixture stirring rate and to extend the reaction time.

4. Organobentonite washing method: with water or with an organic solvent (acetone, ethylene glycol, ethanol and ether). The washing with a solvent excludes an excessive approach and sticking of flakes to one another in the course of drying (or weakens the adhesive forces between them) and provides a high degree of dispersity of the product.

5. Drying method, which determines the degree of organobentonite dispersity. A rapid drying (in a spray drier), where the possibility of agglomeration of particles is avoided, yields a product having a high degree of dispersity, while a slow drying results in agglomeration and coarsening of particles.

The assortment of organobentonite grades commercially produced in various countries and used in production of greases, varnishes, and other products is fairly wide. Thus, Bentone-245 and bentone-structurizer, whose characteristics are presented in Table 3.30, are produced in small amounts in the erstwhile USSR. Bentonite-18c (modifier, octadecylamine), Bentone-34 (modifier, dimethyldiocta-decylammonium halide), Bentone-38, Baragel (modifier, dimethylbenzyloctadecylammonium halide) are produced in the USA and Great Britain; Orben (modifier, mixture of aliphatic ammonium salts and Na-salts of fatty acids), Organite (modifier, quaternary ammonium bases with a benzyl radical in the aliphatic hydrocarbon chain), and Nika Gel L (modifier, quaternary ammonium compounds), in Japan; and Ivegel in Hungary. Other organophilic clays are also produced abroad: Bentone-12, Bentone-27, Nykon-77, which are hectorite treated with ammonium compounds with addition of 30% $NaNO_2$, etc.

Table 3.30. Characteristics of organobentonites produced in the erstwhile USSR

Characteristics	Benton-245, STP 019950 401-031-84 (Askansky deposit)	Bentone-structurized, TU 38.4.01.122-85 (Sarygyukhsky deposit)
Color	Cream	Light-grey to cream
Density, g/cm^3	—	1.6-1.8

Modification degree:
  residual exchange capacity, mg-equiv per 100 g of clay
  mass loss in calcination, %
	30-35
	
	37-40
		not over 10
		
		36-45

Moisture content, %	1.5-3.0	2.0-3.5
Minimum gelation concentration, g per 10 cm^3	—	0.4-0.8
Sedimentation volume of swelling, cm^3/g of bentone	—	18-29

Fractional composition, %:
  passage through screen with mesh No. 014OK GOST 3574-73
  passage through screen with mesh No. 0045K GOST 3574-73
		99.5
		
		99.0

3.3.3. Graphite

Graphite, whose characteristics are given in Chapter 4, after reduction in ball mills and under conditions excluding a contact with oxygen, in presence of organic solvents and other compounds oleophilizes and acquires the ability to form structurized dispersions in organic media [117-122]. This ability of oleophilic graphites is utilized for thickening of various oils with the aim to produce greases.

Graphite is dispursed the medium of an organic liquid whose boiling point is below 500°C and surface is less than 72 N/cm^2 at 25°C in mills of various types. If graphite is oleophilized directly in an oil, the dispersion medium of a grease, it is then mixed with a more volatile organic liquid, which is distilled off on completion of the dispersion. It is recommended to mix petroleum oils with paraffinic hydrocarbon solvents and polar synthetic oils, with esters or alcohols [117-119, 124]. In some cases the oleophilization is conducted in presence of polymers: polyethylene, atactic polypropylene [117-119].

The specific surface of oleophilic graphites is over 20 m^2/g [117-119]. Production of greases requires 5-20% by weight of oleophilic product.

It is proposed to use vermicular (expanded along crystal C-axis) graphite as the thickener [125]. It is produced by impregnation, under stirring, of natural or synthetic graphite with various reagents, such as concentrated H_2SO_4 and HNO_3, their mixtures, sodium or potassiumbichromates, $FeCl_2$, $CrCl_3$, etc., which form layered compounds in contact with graphite. Various degrees of impregnation are attained by varying the conditions (time, temperature, concentration, particle size). After hydrolysis and drying, impregnated graphite is rapidly heated to 600-1000°C (thermal shock). As a result, interlayer bonds get broken and released gases expand the graphite crystal lattice in a direction perpendicular to the plane of layers by a factor over 10, so that particles acquire a vermicular structure.

The bulk density of expanded graphite is 3-5 kg/m^3, and its specific surface as high as 200 m^2/g [126]. A grease thickened with 15% by weight of expanded graphite has a shear strength of 820 Pa at 20°C.

3.3.4. Other Inorganic Thickeners

Technical literature contains propositions of the use, apart from the above-mentioned ones, of other inorganic substances as grease thickeners. These are oxides and hydroxides of various metals, such as hydroxides of alumimium, calcium, barium, strontium and zinc, finely dispersed asbestos, mica, carbonates, sulfides, sulfates, phosphates of metals, etc. [127-137]. These recommendations have found essentially no practical usage and no data on properties of such greases are available. The above listed materials are more often employed as fillers in greases together with soap thickeners.

Use has been found of finely dispersed colloidal asbestos. Having a high specific surface, modified by polymers, aliphatic esters of N-acylaminebenzoic acid in combination with polyethylene, and used as thickener of various oils, makes it possible to obtain waterproof greases, effective at high pressures [150-155]. The percentage of asbestos for thickening of oils varies from 10 to 53 percent by weight.

Asbestos is a chrysolite ($Mg_6(OH)_8Si_3O_{10}$) consisting of separate microfibre of 1-80 mm in diameter and 2,000-30,000 mm in length [130-135]. It can be used, after a treatment similar to that of oleophilic graphite or molybdenum disulfide, not only as an additive, but also as a thickener of lubricating liquids.

3.4. ORGANIC THICKENERS

Various organic compounds exhibiting the ability to form the grease skeleton and of restoring the bonds between particles (structural elements) after their mechanical breakdown as well as good hydrophobicity and anticorrosive properties are also employed for thickening of various lubricating oils to produce greases. The above requirements are met by many organic compounds, but a practical application in grease production has been found primarily by carbon black (commercial carbon), some pigments, polymers and urea derivatives.

We have classified carbon black with organic thickeners because the starting stock for its production are organic compounds, chiefly natural gases and liquid petroleum products.

Main properties of the above-metioned organic thickeners, general notions of their production methods and other data on them are presented below.

Carbon black (commercial carbon). Carbon black has since the 1950's been used as a thickener for grease production [2, 138-141]. It is the cheapest and the most available among organic heat-resistant thickeners. Thus, carbon black, depending on its grade and in respect of the specific consumption of organic thickeners, is 150-300 and 250-600 times cheaper than metal-free phthalocyanine and polytetrafluoroethylene respectively [142].

Carbon black is obtained by an incomplete combustion or thermal decomposition of hydrocarbons. It is one of the structural modifications of carbon and resembles graphite in structure [142-147]. Carbon black has a complex structure, where all carbon atoms forming an elementary particle are bonded together so that the particles are as if a complex polymeric formation consisting of one-, two-, and three-dimensional structures of carbon atoms [143]. Carbon black particles consist of minutest primary crystals, each of which contains 3 to 5 planar crystal lattices of carbon, arranged parallel to one another, and is of a shape close to spherical with a diameter of $(9-400) \times 10^{-6}$ m [144].

Individual spherical carbon black particles are bonded together into linear or branched chains, whose development degree is characterized by the "structureness" (oil number) of carbon black, defined by the amount of linseed oil or dibutylphthalate, absorbed by 1 g of carbon black. This important characteristic of carbon black as a thickener varies from 0.6 cm^3/g for furnace gas carbon black to 1.3-1.6 cm3/g for special carbon black grades. The thickening ability of carbon black is governed not only by its "structure", but also by properties of its surface, which feature high physical, chemical and energetic nonuniformities [142, 144-147]. The physical non uniformity of carbon black surface stems from its roughness, and the chemical one, from a complex chemical composition of the surface particle layer, where oxygen-containing groups—carboxyl, hydroxyl, and carbonyl—are present. Apart from the oxygen-containing groups, the carbon black surface layer contains insignificant amounts of hydrogen, nitrogen, sulfur and other elements, whose presence on the carbon black surface determines its wettability by various liquids. The energetic nonuniformity is characterized by the presence of active sites, higher energy areas, on the surface and, along with the dispersity (specific surface) and elementary composition of the surface, determines its adsorption properties. Depending on the carbon black production method and other factors, the specific adsorption surface of carbon black in terms of nitrogen varies within 60-700 m^2/g; the bulk density, within 25-400 kg/m^3; and the density, within 1800-1950 kg/m^3. Carbon black is resistant to alkalis, acids, and other chemical compounds, little sensitive to light, high and low temperatures [146].

As a thickener used in grease production, carbon black exhibits a universal thickening effect: it thickens equally well hydrocarbon oils, alkylaromatic and silicon liquids, esters and ethers [142]. Carbon black properties (structure, specific surface, adsorption activity, density, bulk density, particle surface state, and other characteristics) depend on its production method and conditions as well as on the starting stock used.

There exist several carbon black production methods, based on an incomplete combustion or thermal decomposition of an organic stock (Table 3.31). Three principal methods are used on a commercial scale: channel, thermal, and furnace methods. Carbon black properties can be varied over a wide range by changing the method and temperature conditions of its production. Commercial carbon (carbon black) is produced according to GOST 7885-77 in the following grades:

Table 3.31. Classification of carbon black and their production methods

Type	Production method	Stock	Subtype and index
Channel	Deposition on metallic surfaces from smoking diffusion flame	Natural gas	For paints (KK) Easily workable (KL) Medium-workable (KS) Hardly workable (KT) Electrically conductive (KE)
Anthracene	Ditto	Raw anthracene, anthracene oil, or naphtalene with coke-oven gas	Channel anthracene (KA)
Furnace	Incomplete combustion in special furnaces or in turbulent gas or air stream	Natural gas or natural gas with addition of liquid hydrocarbons	Semi strengthening (PP) High-module (PM) Disperse (PD) Electrically conductive (PE)
Thermal	Thermal decomposition without access of air in regenerative periodic furnaces	Natural gas or natural gas-air mixture	Thermal (T) Disperse (TD)
Lamp	Combustion of vapors of liquid hydrocarbons evaporated from special bowls by heat of combustion	Green oil and other petroleum residues	Lamp (L)
Atomiser	Combustion of liquid hydrocarbons in flame with stock atomization by atomizers	Green oil and other petroleum residues	Atomizer (F)
Acetylene	Explosive decomposition of acetylene without access of air under elevated pressure Continuous thermal decomposition of acetylene under atmospheric pressure Electrocracking of methane in electric arc with formation of acetylene and then of carbon black	Acetylene Methane	Explosive (AV) Thermal (AT) Arc (AD)

DG-100—*active channel carbon black, produced in a diffused flame by incomplete combustion of natural gas, having a specific geometric surface of about 100 m^2/g and a low "structure" index;*

DMG-80—*active channel carbon black, produced in a diffused flame by incomplete combustion of vapours of hydrocarbon oils in mixture with coke-oven gas, having a specific geometric surface of about 80 m^2/g and a low "structural" index;*

PM-75—*active furnace carbon black, produced by incomplete combustion of hydrocarbon oils, having a specific geometric surface of about 50 m^2/g and a high "structural" index;*

PM-50—*medium-activity furnace carbon black, produced by incomplete combustion of hydrocarbon oils, having a specific geometric surface of about 50 m^2/g and a high "structural" index;*

PGM-33—*semi-active furnace carbon black, produced by incomplete combustion of natural gas, either pure or with addition of hydrocarbon oils, having a specific geometric surface of about 35 m^2/g and a high "structural" index;*

PM-30V—*semi-active furnace carbon black, produced by incomplete combustion of hydrocarbon oils, having a specific geometric surface of about 25 m^2/g and a high "structural" index;*

TG-10—*low-activity thermal carbon black, produced by thermal decomposition of natural gas, having a specific geometric surface of about 15 m^2/g and a low "structural" index.*

Carbon black grades DG-100, DMG-80, PM-75, PM-50, PGM-33, PM-30V and PM-15 are produced in a granulated, and grade TG-10, in a nongranulated form; the physico-chemical properties of these carbon black grades according to GOST 7883-77 are presented in Table 3.32.

In addition, carbon blacks of various grades and purposes are produced to specifications: commercial carbon (TU 14-7-27-80); commercial carbon for manufacture of special products (TU 38-013129-78); commercial carbon PGM-40 (TU 38-11310-77); commercial carbon PM-40N (TU 38-11328-77); commercial carbon DMG-103, DMG-80 (TU 38-11366-77); commercial carbon PM-103K (TU 38-11316-82); special deep-black channel commercial carbon (TU 31-80-82); active furnace commercial carbon PM-103 (TU 38-11362-77); thermal carbon PM-90V (TU 38-11338-73); and carbon black PM-13TS for manufacture of hard-alloy products (GOST 3813.31-73).

CMEA standard 5507-77 (Carbon Blacks, Classification and Terminology) classifies carbon black by the production method, mean arithmetic diameter of particles, specific adsorption surface, "structural" and its other specific properties.

Greases where carbon black is the thickener are called carbon black greases. The experience in their use [2,142] evidences that they can be successfully employed as high-temperature greases. Carbon black is the most available and cheapest thickener for producing thermally stable greases. The optimum properties for production of greases are offered by such commercial carbon blacks where the specific surface is not less than 100 m^2/g and the "structural" is within 0.8-1.2 ml/g, such as DG-100, PM-103, and others [142]. The optimum carbon black content in greases is within 15-22%. It should be taken into account that petroleum and synthetic oil-based carbon black greases are prone to thermal hardening by a prolonged action of high temperatures (200°C). When a thermally stable dispersion medium is used for their production, however, these greases can be used up to 250°C [142,148].

Table 3.32. Physico-chemical characteristics of carbon blacks produced to GOST 7885-77

Characteristics	DG-100	DMG-80	PM-75	PM-50	PGM-33	PM-30V	PM-15	TG-10
Specific geometric surface, m²/g	90-103	80-88	75-82	49-57	—	—	—	—
Specific conventional surface, m²/g	—	—	—	—	32-38	20-26	12-18	12-16
Specific adsorption surface, m²/g	130-160	Not over 95	—	—	—	—	—	—
Oil number, cm³/100 g	—	—	95-105	96-110	60-80	100-120	85-105	—
PH of wter suspension	3.5-4.5	—	7-9	7-9	—	—	—	—
Optical density of gasoline extract (not over)	—	0.3	0.05	—	0.2	0.1	0.1	—
Moisture content, % (not over)	2.5	2.5	0.5	1.0	0.5	0.5	0.5	0.5
Ash content (not over)	0.1	0.1	0.3	0.3	0.5	0.2	0.2	0.2
Total sulfur content, % (not over)	—	—	1.1	1.1	—	—	—	—
Residue after screening through screen with mesh 014OK acc. To GOST 3574-73, % (not over)	0.02	0.03	0.02	0.02	0.02	0.01	0.01	0.02
Bulk density of granulated carbon black, % (not below)	—	—	330	330	400	300	300	—
Dust content in granulated carbon black, % (not over)	—	—	8.0	8.0	8.0	8.0	8.0	—
Abrasion resistance of granules, % (not below)	—	—	87	87	87	87	87	—

Pigments. Natural or synthetic pigments are finely dispersed dyes, most of which feature high chemical and especially thermal stabilities, water-insolubility and structure forming ability in hydrocarbon and other lubricating liquids [149,150]. These unique properties of pigments provided for their use as thickeners for grease production when a rapid progress in the aviation, rocket and other fields of engineering required the development of lubricants serviceable at temperatures of 300-500°C and above [1, 2]. Greases where the thickener is a pigment are called pigment greases (Pg-greases) or dye base greases. Pigments mostly have a relatively low thickening power, and therefore as much as 20-50% thickener is added to oils to produce greases [1, 151]. Pg-greases are generally rated for use upto 250-300°C [1].

The widest use as thickeners for grease production has been found by phthalocyanine pigments, indanthrones, isoviolanthron, indigo and some others [1, 2, 151-158]. Molecules of pigments (phthalocyanines, indanthrones, isoviolanthrons) contain functional groups characteristic of oxidation inhibitors and therefore not only serve as thickeners in greases, but are also additives in upgrading the oxidation and thermal stabilities of the latter [159]. Pigments are particularly efficient for upgrading the thermal stability of polysiloxanes [159] and also enhance the antiwear and antiseizure properties of greases [160].

Phthalocyanines [149, 150] are produced by heating aromatic o-dicarboxylic acids or their anhydrides with ammonia or substances decomposing with release of ammonia (e.g., urea). The reaction can be conducted either in a solvent (trichlorobenzene) or without it, but in presence of catalysts: arsenic pentoxide, ammonium molybdate, boric acid, etc.

Copper phthalocyanine is obtained from phthalonitrile, heating its mixture with CuCl till melting of phthalonitrile (melting point, 141°C), after which an exothermal process (260-300°C) of copper phthalocyanine formation begins:

$$4C_6H_4(CN)_2 + 2CuCl \rightarrow C_{32}H_{16}N_8Cu + CuCl_2.$$

Metal phthalocyanines exist in four crystalline modifications, two α (stable and unstable), β, and γ ones, differing in colour shades and properties. The α- and β-modifications are of practical importance.

Copper phthalocyanine is insoluble in water, oils, alcohols and some other organic solvents, resistant to the action of light, high temperatures, acids and alkalis, and do not decompose in air on heating to 500°C. Copper phthalocyanine is produced under the name of light-blue phthalocyanine pigment (A); it is a uniform light-blue powder. Its empirical formula is $C_{32}H_{16}N_8Cu$; molecular mass, 576.1.

(A)

Individual powder particles or their aggregates are visible in an electron microscope. The vast majority of particles are rectangular, of 0.1-0.2 mm in length, the length-to-width ratio reaching 5. Needle-like particles are encountered less frequently [151].

A metal-free phthalocyanine, containing no coordination-bound metal atom, is obtained from alkali salts of phthalocyanine, usually sodium ones, the salts being produced by interaction of phthalonitrile with alcoholates of these metals in high-boiling alcohols. It is produced under the name of greenish-light-blue phthalocyanine pigment (B). Its empirical formula is $C_{32}H_{18}N_8$; molecular weight, 514.55. The phthalocyanine powder consists of planar rectangular particles of 0.05-0.5 μm in length and with a length-to-width ratio of 1-10. Particles often form 0.3-2 mm long, 0.1-0.6 mm wide agglomerates [151].

Modified stable phthalocyanines should be used in production of phthalocyanine greases since, when metastable forms are employed, the size of their crystals increases with time, especially in petroleum oils and esters and to a lesser extent in poly-siloxanes. It is recommended to subject phthalocyanine pigments to thermal treatment at 220-225°C before their use in order to remove impurities.

(B)

The domestic industry produces more than 10 grades of copper phthalocyanine and several grades of metal-free and chlorinated phthalocyanines. Copper phthalocyanine produced to TU 6-14-516-80 is chiefly used at present in grease production.

Indanthrenes [149] are compounds with condensed anthraquinone rings, derivatives of indanthrone:

The indanthrone molecule is large and planar. Hydrogen bonds between amino groups of the dihydrazine ring and oxygen atoms of carbonyl groups stabilize indanthrone against photooxidation processes. Blue pigments, derivatives of indanthrone are insoluble in organic solvents and acids. They are produced in the erstwhile USSR under names of vat blue "O", blue anthraquinone pigment "SA", and vat light-blue "0" [149,151]. As reported in [151], pigment "SA" is attractive as the thickener for high-temperature greases. It is soluble in strong sulfuric acid. Its empirical formula is $C_{28}H_{18}O_4N_2$ and molecular weight 442.4. Submicronic particles of 0.08 to 0.5 mm in length and 20-30 nm in width are visible in electron micrographs of specimens of an aqueous paste of pigment "SA".

Isoviolanthron is produced in the form of a dry powder or an aqueous paste-suspension to TU 6-14-537-79 which can be used in grease production [151].

ISOVIOLANTHRON

Indigo, one of the earliest organic pigments, is the prototype of vat dyes [149]. It is at present seldom used for these purposes. The structural formula of indigo is

This is a dark-blue water-insoluble powder; it has an amphoteric character, but acidic and basic properties are little marked. Thus, it forms a salt only with concentrated H_2SO_4. It is obtained from aniline and monochloroacetic acid with a subsequent melting of the resulting phenylaminoacetic acid and oxidation of the Na-salt of indoxyl. Indigo is produced to GOST 6392-74 and can be used as the thickener to produce greases.

The research and development efforts on pigment greases in the erstwhile USSR have been conducted for nearly 20 years at the "VNIINP" Institute. The domestic industry produces at present several pigment greases: VNIINP-234 (TU 101435-74), VNIINP-235 (TU 38-101297-78), VNIINP-246 (GOST 18852-73) and grease N 158 (TU 38-101320-77), where the thickener is isoviolanthron or copper phthalocyanine [161].

Polymers. Polymers of various natures and structures have long been used to produce greases for various purposes [1,2]. Some polymers (polysiloxanes, polyfluorohydrocarbons and other liquids) serve as the dispersion medium of greases. Other elastic natural and synthetic high polymers (rubber, polymethacrylates, polyisobutylene, polyolefins, etc.) are employed as additives in lubricating greases. Solid high molecular weight polymeric compounds, mainly those insoluble in hydrocarbon oils and other lubricating liquids, are used as thickeners. There exists a wide variety of these polymers, but they can be divided into two basic groups: fluoropolymers and remaining high molecular weight compounds.

A practical application in production of greases resistant to most chemically corrosive media has been found, by polytetrafluoroethylene, polytrifluoro-chloroethylene, polyfluorophenylene, polyethylene, polypropylene, polybutene and some other polymers [162-173].

Polytetrafluoroethylene (Fluoroplastic-4, Teflon), a carbon-chain polymer—$[—CF_2—CF_2—]_n$, is a solid milk-white product with a molecular weight of 500,000-2,000,000. Its crystal structure is destroyed at about 527°C; then it becomes transparent and goes over into a highly elastic state, which is retained right up to the decomposition temperature. It is produced by polymerization of tetrafluoroethylene in presence of peroxide catalysts [167,173].

The following grades are produced: Fluoroplastic-4 (GOST 10007-80E); Fluoroplastic MB (OST 6-05-400-74); Fluoroplastic-40 (OST 6-05-402-80); and polytetrafluoroethylene (TU 6-05-041-555-74). They are used to manufacture various products in the electrical-engineering, radio and chemical industries. In the grease industry they are used as thickeners in production of chemically and thermally stable special greases SK-2-06 (TU 6-02-786-77) and WIINP-255 (TU 58-101687-77) as well as antifriction additives in greases [161].

Polytetrafluoroethylene-based lubricating greases exhibit high heat resistance, resistance to chemically active substances and good lubricity [164-166]. It should be noted that polytetrafluoroethylene destruction, accompanied by evolution of highly toxic compounds, such as perfluoroisobutylene, whose toxicity is 10 times that of phosgene, occurs on heating to 400-500°C [174].

Polytrifluorochloroethylene (Fluoroplastic-3), a carbon-chain polymer—$[—CF_2—CFC—]_n$, is a solid white product with a molecular weight of 50,000 to 200,000. Its crystallites melt at about 208-210°C, and at 240-270°C it goes over to a viscous-flow state. At low temperatures it does not dissolve and swells very little in solvents; at 130-150°C it dissolves in some aromatic hydrocarbons; it is resistant to acids, oxidants, and alkalis [167]. It is produced by polymerization of trifluorochloro-ethylene in presence of peroxide catalysts. It is used as an anticorrosive structural material for coating of various vessels, pumps, pipes, etc. It can also be employed as the thickener to produce chemically stable greases. Fluorocarbon greases 100 KF (OST 6-02-205-78) and Zf (TU 6-02-796-78), based on this polymer, are produced in this country.

Polyethylene $[—CH_2—CH_2—]_n$, is a carbon-chain polymer. It is commercially produced by several methods: high-pressure, low-pressure, and medium-pressure polymerization.

Different ethylene polymerization mechanisms at different pressures result in formation of polymers differing in structure and properties.

Polyethylene is resistant to various corrosive media. When heated, it swells in hydrocarbons. High-density low-pressure polyethylene (OST 6-20.560.023-73), high-density polyethylene (OST 4 GO.023.055), low-density polyethylene (OST 4 GO.023.056), and medium-density polyethylene (TU 38-10258-81) are produced. Polyethylene has not found wide use as the grease thickener. Greases based on it have a low application temperature, a low adhesion to metal, and "age" rapidly; therefore it is more often employed as thickener.

Polypropylene is a linear carbon-chain polymer with a molecular weight of 30,000-500,000:

$$[—\underset{\underset{CH_3}{|}}{CH}—\underset{\underset{CH_3}{|}}{CH_2}—\underset{\underset{CH_3}{|}}{CH}—CH_2—]_n.$$

The spatial arrangement of side groups relative to the main chain is of crucial importance for polymer properties. There exist isotactic, syndiotactic and atactic propylene polymers [173]. The most important steric variety is the isotactic structure, which exhibits a high degree of crystallinity, high transparence, hardness, and heat stability. The atactic polypropylene is very flexible, soft and sticky, exhibits a high resistance to acids, alkalis and other corrosive media. It is produced by polymerization of propylene in a solvent at a pressure from 1 to 4 MPa and a temperature of about 70°C in presence of catalytic complex $AlCl_3$ + $TiCl_3$. The atactic polypropylene (TU 6-05-1902-81E) and secondary polypropylene (TU 6-19-170-80) are produced. Greases based on them exhibit no high service properties. Their melting points are generally somewhat above 100°C. A polymer with a molecular weight of 50,000-200,000 is used for thickening; its addition in an amount of 10-15% yields greases close to jellies in the appearance [1]. It is more often used as a thickener. Other polymers can also be used as thickening agents and additives in improving functional properties of greases.

Urea derivatives. In the search for greases serviceable over a broad temperature range (particularly with a high upper temperature limit of application), at high specific loads and speeds, in contact with adverse media and offering a high hydrolytic and chemical stability, a good oxidation and mechanical stability, and other valuable performance characteristics, such as a good pumpability, a long service life, it was found that greases thickened with urea derivatives, organic compounds whose molecules contain one or several urea groups, are the most suitable in this respect [1, 175]. These greases were developed nearly at the same time as the pigment greases [1].

Polyurea thickeners and hence also greases based on them can be divided into the following main types [175-193].

Polyurea thickeners, which are a polyurea

$$RNH-\underset{\underset{O}{\|}}{C}-(NH-r-NH-\underset{\underset{O}{\|}}{C}-)_n-(NH-r'-NH-\underset{\underset{O}{\|}}{C}-)_m-(NH-r''-NH-\underset{\underset{O}{\|}}{C})_k-NH-R',$$

where R and R′ → aryl, alkyl, or hydrogen; r, r′, r″—bifunctional aromatic, aliphatic, or heterocyclic radical of a di-isocyanate or a diamine component; and n, m, k—any number, including $n = m = k = 0$. Polyurea thickeners are formed by the interaction of mono- or di-isocyanates with mono- or diamines

$$R-N=CO + R'-NH_2 \rightarrow R-NH-\underset{\underset{O}{\|}}{C}-NH-R'.$$

Isocyanates exhibit a high reactivity; they interact at a high rate in absence of catalysts (except special cases) with amides, amines, alcohols, carboxylic acids, ethers and other compounds, with a mobile hydrogen atom. Aromatic isocyanates are more reactive than aliphatic ones.

Diamines react with di-isocyanates to yield polymers containing reactive amine or isocyanate groups at the ends. To break the chain, they should be blocked by monoisocyanates or monoamines, whose number controls the size (length) of the polyurea molecule.

The following substances for production of polyurea thickeners are mentioned in literature [173-193] isocyanates, *e.g.*, diphenylmethyldiisocyanate, toluenedi-isocyanate bitoluylenedi-isocyanate, octadecyl-isocyanate, hexamethylene di-isocyanate, etc.; and of amines, aminated animal and vegetable fats, octadecylamine, cyclohexylamine, oxydiamine, chloroaniline, toluyilenediamine, chloroamine, phenylenediamine, trifluoromethylaniline, hexamethylenediamine, etc.

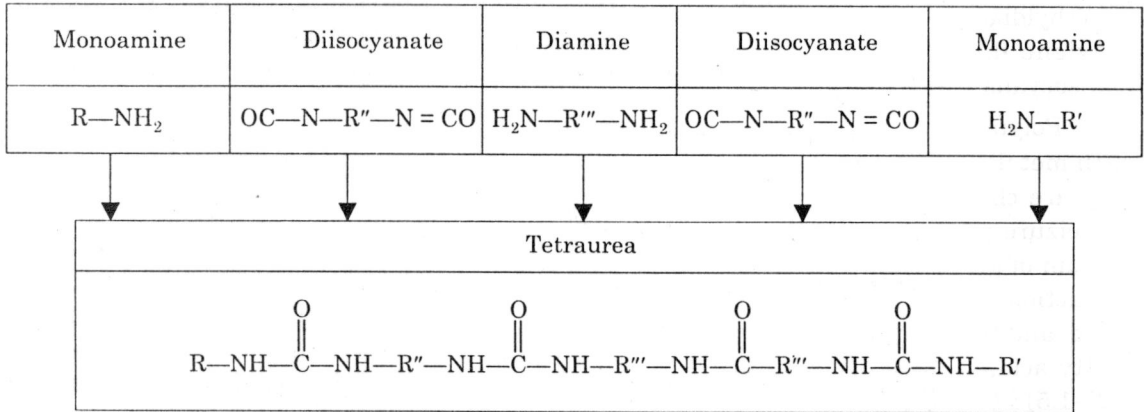

Aryl derivatives of urea are more often used in grease production because its alkyl and acyl derivatives exhibit a poorer thickening power and a lower thermal stability. Most widely used of polyureas is tetraurea, whose production scheme is shown in Fig. 3.5. Such a urea polymer is produced directly in the dispersion medium of the grease at a diamine : diisocyanate : monoamine = 1 : 2 : 2 molar ratio. It is to be noted that this grease type, *i.e.*, based on tetraureas, is the most widespread among polyurea greases.

Metal salts of polyurea (ureides) are obtained by interaction of mono amines, caprolactams, and toluenedi-isocyanates [184], which yields an intermediate that is then treated with an aqueous alkali solution in order to obtain the corresponding salt of this polyurea. The polyurea chain can be extended by using diamine in the reaction.

$$R-NH-\underset{\underset{O}{\parallel}}{C}-NH-\underset{CH_3}{\bigcirc}-NH-C-NH-(CH_2)_5-\underset{\underset{O}{\parallel}}{C}-OH$$

Polyurea urethane thickeners, whose molecules contain urethane ($-NH-\overset{O}{\overset{\parallel}{C}}-O-$) groups [186-189], are produced through interaction of monobasic alcohols with di-isocyanates or of dibasic alcohols with di-isocyanates and monoamines. The general formula of polyurea urethane is

$$R'-O-\overset{O}{\overset{\parallel}{C}}-NH-A'-NH-\overset{O}{\overset{\parallel}{C}}-NH-R''-NH-\overset{O}{\overset{\parallel}{C}}-NH-A''-NH-\overset{O}{\overset{\parallel}{C}}-O-R''',$$

where R' and R''' are aliphatic hydrocarbon radicals containing 4-30 carbon atoms; R'' is aliphatic hydrocarbon radical containing 2-12 carbon atoms; A' and A'' are aromatic hydrocarbon radicals containing 6-18 carbon atoms. To produce this thickener it is expedient to use the following alcohols: nonyl, lauryl, hexadecyl, etc.; diamines: ethylenediamine, propylenediamine, trimethyldiamine, etc.; and diisocyanates: biphenylene diisocyanate, naphthalene diisocyanate, toluylene diisocyanate, 3, 3'-dimethylbiphenyl-4, 4'-diisocyanate, diphenylmethane-4, 4'-diisocyanate, etc.

Polyurea acetate (ureidoacetate) thickeners are a polyurea complex with an alkaline-earth metal salt of a low molecular weight carboxylic acid, mainly calcium acetate [190-192]. They are cheaper than polyurea thickeners and provide for obtaining greases with higher antiseizure properties. It is proposed to produce these thickeners directly in the dispersion medium of greases by the following basic methods: first a polyurea thickener is obtained by interaction of diisocyanates with amines and then calcium acetate or calcium hydroxide is added and then acetic acid is added to the obtained polyurea thickener [190, 191]; polyurea, a fatty acid, such as oleic or stearic acid, and calcium acetate are mixed at a molar ratio (0.25-1.5) : 1 : (0.5-2) [192].

Conventionally put in this group may be an amidoisocyanurate thickener, produced by interaction between fatty acids, such as stearic, C_{10}-C_{20} SFA, ets. and diisocyanate [193] in two stages:

The isocyanatearylamide of fatty acid, thus formed is a poor thickener, and therefore such a component as calcium acetate is added, which, besides is also a catalyst for the above-mentioned process stage and for the second stage, trimerization of isocyanates with formation of the isocyanurate ring. Both process stages are accomplished through reactions proceeding concurrently and in succession. In addition, a part of the amides, interacting with one another, transforms into ureas with separation of acid anhydrides

This thickener is attractive in that it can be produced with the use of relatively cheap and nonscarce stock, such as C_{10}—C_{20} SFA. High-quality greases are produced at a relatively small consumption of polyurea with the use of polyurea acetate complexes.

Dibenzimidazolediureate thickeners are produced by interaction of dicarboxylic acid or of its anhydride with nitro substituted phenylenediamine, followed by reduction of nitro groups into amine groups and a subsequent interaction of the obtained product either with monoisocyanate or with the product of aromatic diisocyanate with monobasic alcohol or monoamine.

The above-mentioned polyurea thickeners can be obtained directly in the grease production process (in situ) or outside the process with the subsequent use of a finished thickener. They are mostly produced in situ.

Despite the high performance characteristics of polyurea greases, they have so far not found widespread use. These greases are produced on a commercial scale in the USA and Japan, and on a pilot scale, in Germany, Poland and some other countries. They were not produced in the former USSR, although the required scope of research and development efforts on production of such greases from domestic stock had been conducted. The factor restraining the production of these greases appears to be the toxicity as well as some scarcity of the stock used to produce polyurea thickeners: isocyanates, amines and other components. Note that, in contrast to the starting stock, neither polyurea thickeners nor greases based on them are toxic; it is, however, not recommended to incinerate their wastes.

Polyureas in a finished state are not produced on a commercial scale in the former USSR. The Dzerzhinsk branch of the GIAP Institute developed a phosgene-free method for production of polymeric ureas of a general formula $[(CH_kR_{2-k})_l$—Q—$(CH_kR_{2-k})_l$—NR'—C—NR'']_m$, where Q is an aromatic, aliphatic, or organosilicon radical; R' and R'' hydrogen or $SiH_n(CH_3)_{3-n}$; $k = 1; 2; l \geq 1; n = 0; 1$; and $m \geq 2$. These compounds can be used as thickeners in production of polyurea based greases. The following products can also be used to produce polyureate thickeners in situ:

(a) 4, 4'-diphenylmethane diisocyanate (4, 4'-MDI), is produced to TU 113-03-29-3-80 in the form of white (highest grade) or yellow (first grade) flakes which at 38°C convert into liquid with a density of 1190 kg/m³ at 40°C. It is toxic and combustible (flash point, 216°C). Safety precautions set forth in the TU and departmental instructions should be observed in handling it;

(b) Commercial phthalolamine (octadecylamine of stearic acid), produced to GOST 25717-79 is a wax-like mass of white or light-grey color with a flash point of 167.3°C in an open crucible, an ignition point of 188°C and a self-ignition point of 280°C. It belongs to a class of moderately hazardous compounds, exerts a harmful action on the skin and mucous membranes of eyes; the MPC of its vapour in the air of the working zone is of 1 mg/m³;

(c) Benzyl amine (TU 38.4042-83), produced by hydrogenation of benzonitrile. It is a colourless or slightly yellowish liquid containing 98.5% basic substance; its flash point is 75°C and ignition point is 80°C. It belongs to moderately toxic products; the MPC of its vapor in the working zone air is 2 mg/m³. Handling of phthalolamine and benzyl amine as well requires observing the corresponding safety rules.

Other organic compounds. Fluorinated carbon black can be used as a thickener of lubricating liquids, producing greases with excellent lubricating properties. It is produced by action of elemental fluorine on carbon black at 400-600°C in a nitrogen or an argon atmosphere [195] by the reaction: $nC + xnF \rightarrow (CF_x)_n$.

The fluorination is carried out in a reactor, where carbon black powder is activated at 250-300°C and blown through with nitrogen to remove volatile impurities, after which a fluorine-nitrogen mixture at 400-440°C is fed for 16-20 h [195]. On completion of the fluorination the reactor is cooled to 200°C, purged with nitrogen and unloaded. The carbon-to-fluorine ratio in fluorinated carbon black varies from 0.25 to 1.0 and above. A product with high lubricating characteristics is obtained at a C : F ratio above 1.0.

Greases based on fluorinated carbon black DG-100 offer a number of advantages over greases thickened by ordinary carbon black DG-100: an upgraded mechanical stability, a higher thermomechanical stability, a 17-33-% lower wear of rubbing surfaces, and an 1.5-2 times higher serviceability in the friction zone at 250°C [196].

Ammeline, diamide of cyanuric acid (2,4-diamine-6-hydroxy-1,3,5-simtriazine), can also be used as a thickener to produce high temperature greases exhibiting high performance characteristics: serviceability in vacuum, at high rotation speeds, considerable axial and radial loads, etc. [197-199]. It was ascertained [198] that the thickening effect of ammeline depends on the method of its production and isolation from the reaction mass. The optimum ammeline production method is the process consisting of the following stages: incomplete amination of cyanuric chloride; saponification of the intermediate product with alkali; neutralization of the reaction mass with acid; filtration and drying. The resulting product is a white odourless solid substance containing 85-88% basic product, having a specific surface in terms of nitrogen of 12-14 m^2/g, a pore volume of about 0.6 cm3/g, a bulk density of 400-600 kg/m3 and pH of aqueous extract 7-8 and containing: ash not over 0.3%; water, upto 3%, and no chloride and iron ions.

Ammeline not only offers a good thickening power, but also serves as an oxidation inhibitor as well as enhances antiwear and antiseizure properties of the dispersion medium [198-200].

Sodium amidoterephthalate is also used as a thickening component of high temperature greases offering a high mechanical stability, water resistance and other valuable properties [201].

Other organic substances having an extensive specific surface and a thickening effect in petroleum oils and other lubricating liquids can also be used as thickeners in production of greases.

REFERENCES

1. Sinitsyn V.V., The Choice and Application of Plastic Lubricating Greases, Khimiya, Moscow, 1974.
2. Boner K.D., Production and Application of Greases. Transl. from English, Ed. by Sinitsyn V.V., Gostoptekhizdat, Moscow, 1958.
3. Ishchuk Yu. L. and Oherednichenko G.I., Production and Application of Greases and Prospects for Their Advance, in: Plastic Greases, Proc. of 2nd All-USSR Sci. and Techn. Conf. (Berdyansk, 1975), Naukova Dumka, Kiev, 1975, pp. 5-7.
4. Ishchuk Yu. L., State and Prospects of Production and Application of Plastic Lubricating Greases, Neftepererabotka i Neftekhimiya, 1977, N 10, pp. 54-56.
5. Chemist's Reference Book, Khimiya, Moscow-Leningrad, 1964, Vol. 2.
6. Nekrasov B.V., Fundamentals of General Chemistry, Khimiya, Moscow, 1973, Vol. 2.
7. Ishchuk Yu.L., Influence of Dispersed Phase Composition on Structure and., Properties of Oleogels, Plastic Lubricating Greases, in: Physico-Ghemical Mechanics and Lyophilicity of Disperse Systems, Naukova Dumka, Kiev, 1983, Issue 15, pp. 63-75.
8. Reference Book on Soapmaking, Pishchevaya Promyshlennost, Moscow, 1974.
9. Bezzubov L.P., Chemistry of Fats, Pishchevaya Promyshlennost, Moscow, 1974.
10. Tyutyunnikov V.K., Haumenko P.V., Tovbin I.M. and Faniev G.G., Fat Processing Technology, Pishchevaya Promyshlennost, Moscow, 1970.
11. Tyutyunnikov V.K., Chemistry of Fats, Pishchevaya Promyshlennost, Moscow, 1974.
12. Mankovskaya N.K., Chemistry and Technology of Fatty Stock for Plastic Lubricating Greases, Khimiya, Moscow, 1976.
13. Belkevich P.I. and Golovanov V.G., Wax and Its Commercial Analogs, Nauka i Tekhnika, Minsk, 1980.
14. Mankovskaya N.K., Synthetic Fatty Acids, Khimiya, Moscow, 1965.

15. Ishchuk Yu. L., Investigation into Influence of Dispersed Phase on Structure, Properties, and Technology of Plastic Lubricating Greases, Dr. Tech. Sci. Thesis, Moscow, 1978.
16. Mankovskaya N.K., Maskaev A.K. and Krasnova S.I., Monooxystearic Acids: Stock for High-Quality Plastic Lubricating Greases, Naukova Dumka, Kiev, 1971.
17. Pavlikova E.N., Kobylinskaya I.E., Dracheva S.I. and Lendyel I.V., Effect of Thermal Treatment on Quality of 12-Oxystearic Acid, Maslozhirovaya Promyshlennost, 1984, N 3, pp. 32-33.
18. Niyazov A.N., Naphthenic Acids, Ylym, Ashkhabad, 1969.
19. Petroleum Products: Properties, Quality, Application (Reference Book), Khimiya, Moscow, 1966.
20. Trapeznikov A.A. and Zakieva S.Kh., Temperature Dependence of Limiting Shear Stress of Models of Soap Greases, Kolloidn. Zhurn., 1946, Vol. 8, N 6, pp. 429-438.
21. Inventor's Certificate 210, 309 (USSR). Method for Production of Thickener for Plastic Lubricating Grease. Ishchuk Yu. L., Nakonechnaya M.B. and Sinitsyn V.V. Publ. in B.I., 1968, N 6.
22. Daurov B.K., Thermal Stability of Salts of Higher Fatty Acids, Used in Petrochemical Industry, Cand. Tech. Sci. Thesis, Kiev, 1968.
23. Polishuk A.T., Physical and Chemical Properties of Complex Soap Greases, Lubricat. Eng., 1963, Vol. 19, N 2, pp. 76-87.
24. Mankovskaya N.K., Nikishina Z.M. and Maksimilian A.P., Production of Aluminium Soaps of Synthetic Fatty Acids, Neftepererabotka i Neftekhimiya, 1973, N 3, pp. 20-21.
25. Vinogradov G.V., Phase Transitions in Alkaline-Earth Metals, Dokl. AN SSSR, W7, Vol. 58, N 1, pp. 73-75.
26. Void M.J. and Void R.D., A Comparative Study of the X-ray Diffraction Patterns and Thermal Transitions of Metal Soaps, J. Amer, Oil Chem. Soc., 1949, Vol. 26, N 10, PP. 520-525.
27. Vinogradov G.V., Soaps, Their Solutions and Gels, Uspekhi Khimii, 1951, Vol. 20, N 5, pp. 533-559.
28. Dagaev V.A., Buyanova Z.M. and Tanchuk Yu.V., Differential Thermal Analysis of Alkaline Soaps of Monooxystearic Acids and of Greases on Their Base, in: Plastic Greases, Proc. of 2nd All-USSR Sci. and Techn. Conf. (Berdyansk, 1975), Naukova Dumka, Kiev, 1975, pp. 137-138.
29. Buyanova Z.I., Dagaeva V.A., Maskaev A.K. et al., Thermal Analysis of Alkaline-Earth Soaps of Oxystearic Acids, Thickeners of Lubricants, Zhurn. Prikl. Khimii, 1977, Vol. 50, N 6, pp. 1353-1357.
30. Dagaev V.A., Lozovaya V.I. and Mankovskaya N.K., Phase transitions in Aluminium Soaps and Soap-Hydrocarbon Systems, Neftepererabotka i Neftekhimiya, Kiev, 1976, Issue 14, pp. 46-49.
31. Differential Thermal Analysis of Lithium Soaps and Plastic Lubricating Greases, Neftepererabotka i Neftekhimiya, All-USSR Association "Neftekhim", 1976, Issue 11, pp. 17-27.
32. Volobuev N.K., Zubkevich A.V., Ataeva O.V. et al., Differential Thermal Analysis of Commercial Thickeners of Plastic Lubricating Greases, in: Plastic Lubricating Greases, Moscow, 1976, pp. 85-88 (Proc. VNIINP, Issue 16).
33. Primak R.G., Kanchenko Yu.A. and Mysak A.E., Influence of Method of Production of Lithium Soaps on Their Phase Behavior, Neftepererabotka i Neftekhimiya, Kiev, 1978, Issue 16, pp. 12-15.
34. Novoded R.D., Krasnokutskaya M.E., Mysak A.E. et al., Phase Transitions of Calcium and Lithium Soaps of 12-0:xystearic Acid, Khimiya i Tekhnologiya Topliv i Masel, 1985, N 5, pp. 28-29.
35. Vold R.D., Grandine J.D. and Void M.I., Polymorphic Transformations of Calcium Stearate and Calcium Stearate Monohydrate, J. Col. Sci., 1948, Vol. 3, N 3, pp. 339-361.

36. Krasnokutskaya M.E., Godun B.A., Nakonechnaya M.B. and Tikho-nenko M.I., Phase Transitions in Calcium Soaps of Stearic and 12-Oxystearic Acids, in: Plastic Greases, Abstr. of Reports at 3rd All-USSR Sci. and Techn. Conf. (Berdyansk, Sept., 1979), Naukova DmDka, Kiev, 1979, PP. 56-58.
37. Hattiangdi G.S., Vold M.J. and Void R.D., Differential Thermal Analysis of Metal Soaps, Ind. Eng. Chem., 1949, Vol. 41, N 10, pp. 2320-2324.
38. Shekhter Yu. N., Bogdanova T.I., Teterina L.N. et al., Polarity and Functi-onal Properties of Stearic Acid Soaps, Khimiya i Tekhnologiya Topliv i Masel, 1975, N 9, pp. 56-58.
39. Shekhter Yu. N., Fuks I.G., Teterina L.N. et al., On Effect of Polarity of Soaps and Micellar Energy on Structure and. Properties of Plastic Lubricating Greases, Ibid., 1970, N 12, pp. 23-26.
40. Levant G.E., Some Colloidal Properties of Soap-Mineral Oil System, in: Theory and Practice of Production and Application of Greases, Ed. by Velikhovsky D.S., Moscow, 1939, pp. 51-92.
41. Shinoda K., Nakagawa T., Tamamushi B. and Ishemura T., Colloidal Surfactants. Transl. into Russian, Mir, Moscow, 1966.
42. Shibryaev S.B., Fuks I.G. and Tikhonov V.P., Cooling Regime and Structurization of Lithium Greases in Presence of Stearic Acid, Khimiya i Tekhnologiya Topliv i Masel, 1981, N 2, pp. 54-56.
43. Tikhonov V.P., Fuks G.I., Shibryaev S.B. and Fuks I.G., On Effect of Fatty Acids on Lithium Stearate Association in Nonpolar Liquid, Kolloidn. Zhurn., 1980, Vol. 42, N 5, pp. 906-910.
44. Shekhter Yu.N., Krein S.E. and Teterina L.N., Oil-Soluble Surfactants, Khimiya, Moscow, 1978.
45. Demchenko P.A., Scientific Fundamentals of Preparation of Surfactant Compositions, Zhurn. Vses. Obshch. im. D.I.Mendeleeva, 1966, Vol. 11, N 4, pp. 381-387.
46. Ishchuk Yu.L., Kuzmichev S.P., Krasnokutskaya M.E. et al., The Present State and Future Prospects for Advances in Production and Application of Anhydrous and Complex Calcium Greases, TsNIITEneftekhim, Moscow, 1980.
47. Chernozhukov N.I., Petroleum and Gas Processing Technology. Textbook for College Students, Ed. by Gureev A.A., Bondarenko B.I., Khimiya, Moscow, 1978, Pt.3. Purification and Separation of Petroleum Stock, Production of Commercial Petroleum Products.
48. Bogdanov N.F. and Pereversev A.N., Deparaffination of Petroleum Products, Gostoptekhizdat, Moscow, 1961.
49. Goldberg D.O. and Krein S.E., Lubricating Oils from Eastern Petroleum Fields, Khimiya, Moscow, 1972.
50. Pereversev A.K., Bogdanov N.F. and Roshchin Yu.N., Production of Paraffins, Khimiya, Moscow, 1973.
51. Griazov G.I. and Fuks I.G., Production of Petroleum Oils, Khimiya, Moscow, 1976.
52. Petroleum Products: Fuels, Bitumens, Paraffins, Solvents. Collection of USSR Standards, Publ. Board of Committee for Standards, Measures, and Measuring Instruments, Moscow, 1970.
53. Kukovitsky M.M., Nedogrei P.M., Experience in Paraffin and Ceresin Production from Eastern Petroleums, GOSINTI, Moscow, 1959.
54. Sergienko S.R., High-Molecular Petroleum Compounds, Gostopteklhizdat, Moscow, 1964.
55. Gun R.B., Petroleum Bitumens, Knimiya, Moscow, 1973.
56. Baranovsky N..F. and Sukhareva M.P., Ozokerites, Gostoptekhizdat, Moscow, 1959.
57. Investigation of Ozokerites and Ceresins of Various Origins, Gostoptekhizdat, Moscow, 1960 (Proc. of VNIINP, Issue 27).

58. Ivanovsky L., Encyclopaedia of Waxes, in 2 Vol., Gostoptekhizdat, Leningrad, 1956. Vol. 1. Waxes and Their Main Properties.
59. Golovanov N.G., Investigation of Resinous Part of Bitumens of Solid Combustible Minerals, Cand. Chem. Sci. Thesis, Moscow, 1970.
60. Rode V.V., Zharova M.N. and Kostyuk V.A., Main Problems of Obtaining and Using Lignite Wax, Khimiya Tverdogo Topliva, 1974, N 6, pp. 105-108.
61. Belkevich P.I., Gaiduk K.A., Zuev T.T. et al., Peat Wax and Accompanying Products, Kauka i Tekhnika, Minsk, 1977.
62. Nedbailyuk N.S., Research and Development of Hydrocarbon Rope Greases Serviceable over Broad Temperature Range, Cand. Chem. Sci. Thesis, Moscow, 1975.
63. Nikolina V.Ya. and Khysh N.A., Production of Highly Disperse Silica by Vapour-Phase Hydrolysis Method, in: Efforts on Production Technology of Mineral-Origin Fillers and Adsorbents, GKhI, Leningrad, 1963, pp. 64-75 (Proc. of NIOKhim, Vol. 15).
64. Manchenko A.V., Aerosil, Its Properties, Application, and Specifications, Kamenyar, Lvov, 1965.
65. Gurenkov B.S., Makeeva E.D., Nikolaeva I.N. and Nikitin Yu.S., Greases Produced by Thickening of Oils by Silica Gel and Alumino-silica Gel, in: Lubricants, Additives for Them, and Paraffins, Gostoptekhizdat, Moscow, 1958 (Proc. of VNIINP, Issue 17).
66. Bakaleinikov M.B., Plastic Lubricating Grease Production Plants, Khimiya, Moscow, 1977.
67. Chernozhukov N.I., Petroleum and Gas Technology, Khimiya, Moscow, 1978, Vol. 3., Ch. 10. Production and Properties of Plastic Lubricating Greases.
68. Chuiko A.A., SiOp Surface Chemistry, Nature and Role of Active Sites of Silica in Adsorption and Ghemisorption Processes, Dr. Chem. Sci. Thesis, Kiev, 1971.
69. Mashchenko V.I., Chemisorption Sites of Disperse Silica in Reactions with Methylchlorosilanes, Cand. Chem. Sci. Thesis, Kiev, 1971.
70. Sobolev V.A., Quantitative Study of Surface of Highly Disperse Silicas-Aerosils by IR Spectroscopy, Oand. Chem. Sci. Thesis, Kiev, 1971.
71. Pat. 2,657,149 (USA). Method of Esterifying the Surf of a Silica Substrate Having a Reactive Silanol Surface and Product Thereof. K. Ralph Iler, Brandywine Hundred. Publ. Oct. 27, 1953.
72. Mashchenko V.M., Ishchuk Yu.L., Khaber N.V. et al., Butoxy-aerosils: Thickeners of Lubricants, Khimiya i Tekhnologiya Topliv i Masel, 1972, N 1, pp. 55-58.
73. Ishchuk Yu.L., Umanskaya O.I., Makarov A.S. et al., Dependence of Properties of Plastic Lubricating Greases on State of Surface of Silica, Modified toy n-Butyl Alcohol, Ibid., 1977, N 12, pp. 44-47.
74. Chmura M., Badanie zaieznosci miedzu wlasnosciami krzemionek o ich przydatnosci do produkcji smarov, Nafta, 1979, Vol. 35, N 8-9, pp. 295-500.
75. Umanskaya O.I., Physico-Chemical and Rheologic Studies of Plastic Lubricants Based on Modified Aerosil, Cand. One. Sci. Thesis, Kiev, 1974.
76. Bakaleinikov M.B., Investigation in Pield of Silica Gel Plastic Lubricating Greases, Cand. Tech. Sci. Thesis, Moscow, 1974.
77. Ishchuk Yu.L., Umanskaya O.I., Yaniv A.A., et al., Properties of Organo-aerosil-Based Plastic Lubricating Greases with Imine Groups in Modifying Layer, Khimiya i Tekhnologiya Topliv i Masel, 1978, N 8, pp. 57-58.
78. Pat. 2,573,650 (USA). Water-Resistant Greases. W.H.Peterson. Publ. March 22, 1949.

79. Umanskaya O.I., Rudakova N.Ya., Ishchuk Yu.L.. et al., Effect of Chemical Modification of Aerosil hy Dimethylchlorosilane on Grease Properties, Neftyanaya i Gazovaya Promyshlennost, 1975, N 5, pp. 28-50.
80. Pat. 3,526,594 (USA). Grease Composition. S. Meghir. Publ. Sept. 1, 1970.
81. Pat. 3,037,933. Greases Made by Thickening Silicone Oils with Silica and Silica Coated with Octamethylcyclotetrasil Oxane. J.H.Wright. Publ. Oct. 51, 1960.
82. Voloshchuk L.P., Kukharskaya E.V., Skorik Yu. I., Organopoly-siloxy-aerosil as Thickener of Plastic Lubricating Greases, in: Production of Plastic Lubricating Greases and Improvement.
83. Umanskaya O.I., Sheremeta B.K., Chuiko A.A. et al., Dependence of Properties of Aerosil Greases on Thickener Surface Modification Degree, Kef tepererabo flea i Neftekhimiya, Kiev, 1972, Issue 10, pp. 7-11.
84. Yaniv A.A., Ishchuk Yu.L., Umanskaya O.I. et al., Aerosil Plastic Lubricating Greases Based on Synthetic Lubricating Liquids, Neftyanaya i Gazovaya Promyshlennost, 1974, N 5, pp. 39-41.
85. Pat. 3,344,066 (USA). Silicone Greases of High Mechanical and Thermal Stability. H.M.Schelfer, J.W.Vandyke!, J.S.Delphenich. Publ. April 14, 1966.
86. Pat. 2,980,612 (USA). Water-Resistant Non-Soap Greases. R.A.Potter. Publ. Aug. 12, 1957.
87. Christian J.B., Depolarization of Grease Thickener Solids by Polymer Adsorption, NLGI Spokesman, 1971, Vol. 35, N 6, pp. 188-190.
88. Kukharskaya E.V., Voloshchuk A.A. and Voznesenskaya M.M., Modified Aerosil as Thickener of Plastic Lubricating Greases, Khimiya i Tekhnologiya Topliv i Masel, 1977, N 1, pp. 37-40.
89. Krugly I.M., Makeeva E.D., Veismau S.G. and Mikhailov K.M», Bentone Greases as Substitutes of Solidol- and Konstalin-Type Greases, in: Lubricants, Additives for Them, and Paraffins, Gostoptekhizdat, Moscow, 1958 (Proc. of VNIINP, Issue 7).
90. Crookshank F.T. and Bright G.S., The Evaluation of Components for Nonsoap Thickened Greases, NLGI Spokesman, 1977, Vol. 40, N 12, pp. 422-427.
91. Jarocki B. and Gut A., Smary bentonitowe, Nafta, 1972, Vol. 28, N 8, pp. 307-372.
92. Haden W.L. and Schwint I.D., Attapulgite - Its Properties and Application, NLGI Spokesman, 1968, Vol. 31, N 12, pp. 432-440.
93. Kukharskaya E.V. , Vishnevskaya N.P. and Oshueva N.A., Study of Possibility for Using Domestic-Deposit Vermiculites in Lubricating Compositions, Deposited Manusripts, VINITI, Series of Natur. and Exact Sci. and Technol., 1978, No. 4, p. 138. Manuscript deposited at TsNIITEneftekhim, N 11d-454.
94. Ovcharenko F.D., Kirichenko N.T., Kovalenko D.N. et al., Ukrainian Bentonites, Izd. AN USSR, Kiev, 1938.
95. Ovcharenko F.D., Tarasevich Yu.I., Radul N.N. et al., Study on Natural Sorbents of Various Crystalline Structures, In: Natural Sorbents, Nauka, Moscow, 1967, pp. 25-45.
96. Tarasevich Yu.I. and Ovcharenko F.D. , Adsorption on Clay Minerals, Naukova Dumka, Kiev, 1975.
97. Merabishvili M.S. and Magebeli G.A., Bentonites of Soviet Union as Stock for Production of Mineral Sorbents, In: Natural Sorbents, Nauka, Moscow, 1967, pp. 218-223.
98. Vdovenko N.V., Moraru V.N. and Ilyinskaya G.G., Electron Microscope Studies of Micro structure of Organoclays, In: Physico-Chemical Mechanics and Lyophilicity of Disperse Systems, Naukova Dumka, Kiev, 1971, Issue 3, pp. 27-50.

99. Tokmakov V.P., Stock Resources of Vermiculite in the USSR, In: Natural Mineral Fillers (Their Resources and Industrial Use), Proc. of Inst. of Ore Geology, Deposits, Petrography, Geology, and Geochemistry, 1963, Issue 95, pp. 71-75.
100. Ovcharenko V.D., Vdovenko N.V. and Moraru V.N., Effect of Nature of Surfactants on Colloid-Chemical Properties of Disperse Minerals, In: Physico-Chemical Fundamentals of Surfactant Application, FAN, Tashkent, 1977, pp. 69-73.
101. Vdovenko N.V., Zhukova A.I., Dyachenko N.S. and Vasilyev N.G., Kinetics of Ion-Exchange Interaction of Butylammonium Chloride with Cation-Substituted Forms of Montmorillonite, In: Physico-Chemical Mechanics and Lyophilicity of Disperse Systems, Naukova Dumka, Kiev, 1971, Issue 3, pp. 14-21.
102. Sharkina E.V., Structure and Properties of Organomineral Compounds, Naukova Dumka, Kiev, 1976.
103. Ovcharenko F.D., Moraru V.N., Vdovenko N.A. et al., Adsorption Pro-perties of Organosubstituted Vermiculite, Ukr. Khim. Zhurn., 1969, Vol. 35, pp. 269-273.
104. Ishchuk Yu.L., Umanskaya O.I., Yaniv A.A. et al., Effect of Nature of Surface of Organoderivatives of Clay Minerals on Their Structurization in Lubricating Liquids, Kolloidn. Zhurn., 1979, Vol. 41, N 3, pp. 428-434.
105. Kubic C. and Mitterhauszer F., Bentony na pripravu plasti-ckych maziv, Ropa a Uhlie, 1966, Vol. 8, N 7, pp. 211-215.
106. Brindley G.W., Clay - Organ. Stud. USA.—A Review: Abstr. U.S.-Jap. Sem. Clay - Organ. Comp., 1971, pp. 1-4.
107. Suito E., Review of Researches on Clay-Organic Completes in Japan: Abstr. U.S., Ibid., pp. 5-21.
108. Chuzo-Kato, Industrial Application of Clay-Organic Complexes: Abstr. U.S., Ibid., pp. 105-109.
109. House R.F., A Modified Clay Thickener for Corrosion-Resistant Grease, NLGI Spokesman, 1966, Vol. 30, N 1, pp. 11-17.
110. Fariss R.F., A Modified Clay Thickener for Lubricating Fluids, Ibid., 1960, Vol. 23, N 11, pp. 432-437.
111. Pat. 3,222,279 (USA). Lubricant Compositions. E.Loettler. Publ. Dec. 7, 1965.
112. Pat. 3,036,001 (USA). Preparation of Clay Thickened Grease. D.E.Zolffer. Publ. May 22, 1962.
113. Pat. 2,981,685 (USA). Lubricating Oil Thickened to a Grease with Asphaltene Absorbed Clay. D.E. Loeffler. Publ. Apr. 25, 1961.
114. Hydrophobing Agents: Patents and Developments, NLGI Spokesman, 1956, Vol. 20, N 7, pp. 54-44.
115. Ovcharenko P.D. and Gudovich N.V., Mineral Fillers of Rut-tiers, Khim. Promyshlennost, 1960, N 4, pp. 17-19.
116. Vdovenko N.V., Pokhodnya E.A., Modified Kaolin—Rubber Strengthener, Ibid., 1964, N 2, pp. 9-11.
117. Pat. 1,214,985 (Great Britain). Improved Graphite Compositions and Lubricating Compositions Containing Them. A.J. Croszek. Publ. Dec. 9, 1970.
118. Pat. 1,528,491 (Great Britain). Grinding Process. W.J.Alexander, D.S. Ward, K.J. Ware. Publ. Aug. 50, 1975.
119. Pat. 1,575,852 (Great Britain). Oleophilic Graphite Grease. S. Gh. Dodson, K.J. Hole. Publ. Nov. 27, 1974.
120. Pat. 5,770,654 (USA). Grease Composition. S. Ch. Dodson, K.J.Hole. Publ. Nov. 6, 1975.

121. Pat. 1,605,294 (France). Precede de production d'un compose oleophilic modibie et produits obtenus. Hadeuf. Publ. Apr. 5, 1974.
122. Pat. 291,248 (Austria). Verfahren zur Herstellung einer Schmiermittel-komposition. Publ. May 11, 1970.
123. Pat. 1,374,976 (Great Britain). Oleophilic Graphite. S. Ch. Dodson, K.J.Hole. Publ. Nov. 20, 1974.
124. Pat. 2,158,944 (FRG). Verfahren zur Herstellung von oleophilen Graphit und Verwendung des Graphits in einem Schmierfett. K. Schonwald et al. Publ. July 20, 1972.
125. Inventor's Certificate 896,060 (USSR). Thickener of Petroleum Oils. V.A.Filatov, I.L. Maryasin, A.N. Antonov et al. Publ. in B.I., 1982, N 1.
126. Butyrin G.M., High-Porous Carbon Materials, Khimiya, Moscow, 1976.
127. Pat. 2,594,822 (USA). Preparation of Grease. F.H. Stress, S.T. Abrams, R.J. Moore. Publ. Apr. 29, 1952.
128. Pat. 2,599,683 (USA). Grease Composition. S.T. Abrams, F.H. Stross. Publ. June 10, 1952.
129. Pat. 1,039,753 (Great Britain). Improvements in Greases. J. Blowes. Publ. Aug. 24, 1966.
130. Pat. 3,424,678 (USA). Lubricant Containing Alkaline-Earth Metal Mixed Salt Thickeners and Colloidal Asbestos. A.J. Morway, A.J. Bodner. Publ. Jan. 28, 1969.
131. Pat. 3,639,236 (USA). Colloidal Asbestos-Complex Aluminium Salt-Alkali-Alkaline Earth Metal Mixed Salt / Soap Lubricant. A.J. Morway, A.J. Bodner. Publ. Febr. 1, 1972.
132. Pat. 3,639,644 (USA). Colloidal Asbestos Polyethylene Grease. A.J. Morway, A.J. Bodner. Publ. Febr. 1, 1972.
133. Pat. 3,773,663. Greases Thickened with Fibrous Asbestos which Fibers Are Coated with Aliphatic Esters on N-Acyl (Aliphatic) Aminobenzoic Acid. E.A. Cross, R.I. Frye, R.T. Daniel, G.S. Bright. Publ. Nov. 20, 1973.
134. Pat. 3,846,312 (USA). Grease Thickened with Coated Fibrous Asbestos. E.A. Cross, R.I. Frye. Publ. Nov. 5, 1974.
135. Pat. 1,215,271 (Great Britain). Colloidal Asbestos Grease. A.I. Morway, A.J. Bodner. Pabl. Dec. 9, 1970.
136. Pat. 1,282,875 (Great Britain). Oleophilic Metal Sulphides. A.J. Groszek. Publ. July 26, 1972.
137. Pat. 3,361,665 (USA). Metal Phosphate Thickening Agent and Compositions Containing Such. K. Tesche, K. Griessberger, W. Lamper. Publ. June 2, 1968.
138. Pat. 2,467,148 (USA). Oil Insoluble Lubricant. A.J. Morway, D.V. Young, Publ. Apr. 12, 1949.
139. Pat. 2,467,145 (USA). Lubricant. A.J. Morway, J.C. Zimmer. Publ. Apr. 12, 1949.
140. Pat. 2,467,146 (USA). Lubricant. A.J. Morway, J.C. Zimmer. Publ. Apr. 12, 1949.
141. Pat. 2,467,147 (USA). Low-Temperature Lubricant. A.J. Morway, J.C. Zimmer. Publ. Apr. 12, 1949.
142. Shulzhenko I.V., Kobzova R.I., Kikonorov E.M. and Bakalei-nikov M.B., Carbon Black Plastic Lubricating Greases (Review). TsNIITEneftekhim, Moscow, 1983.
143. Lezhnev N.N., Krasilnikov M.K. and Rybakova V.I., Thermome-chanical Study of Carbon Blacks, In; Production and Properties of Carbon Blacks, Zap.-Sib. Kh. Izd., Omsk Old., Omsk, 1972, pp. 49-55.
144. Belenky E.F. and Riskin I.V., Chemistry and Technology of Pigments, Goskhimizdat, Leningrad, 1960.

145. Syunyaev Z.I., Petroleum Carbon, Khimiya, Moscow, 1980.
146. Kaltsev V.V. and Tesner P.A., Carbon Black. Properties, Production and Application, Gostoptekhizdat, Moscow, 1952.
147. Zuev V.P. and Mikhailov V.V., Production of Carbon Black, Khimiya, Moscow, 1965.
148. Shulzhenko I.V., Kobzova R.I. and Oparina E.M., Carbon Black Plastic Lubricating Greases, In: Plastic Lubricating Greases, Khimiya, Moscow, 1976, pp. 60-66 (Proc. of VHIINP, Issue 16).
149. Stepanov B.I., Introduction to Chemistry and Technology of Organic Dyes, Khimiya, Moscow, 1977.
150. Moser F.H. and Thomas A.J., Phtalocyanine Compounds, Reinhold, New York, 1969.
151. Oparina E.M., Vselyubsky S.B., Dmitrieva V.G. et al., High-Temperature Greases, In: Plastic Lubricating Greases and Solid Lubricating Coatings, Khimiya, Moscow, 1969, PP. 164-183 (Proc. of VKTIIKP, Issue 11).
152. Pat. 2,597,018 (USA). Lubricant. R.L. Merker, C.R. Single-terry. Publ. May 20, 1952.
153. Pat. 2,929,779 (USA). Silicone Oil Grease Containing Phtalocyanine and Acetylene Black. F.F. Sullivan, H.R.Baker. Publ. March 22, 1960.
154. Pat. 906,009 (Great Britain). A High-Temp era lure-Resist ing Lubricant Based, on Polysiloxane. Wacker-Chemie GmBH, Publ. Sept. 19, 1962.
155. Pat. 1,123,421 (FRG). Hochtemperaturschmierfett. D.E. Loef-fler. Publ. Aug. 30, 1962.
156. Pat. 802,896 (Great Britain). Lubricating Grease Compositions. N.Y. DC. Botaafsche Maatschapii. Publ. Oct. 15, 1958.
157. Inventor's Certificate 131,432 (USSR). Anti-Rust Oil. A.N. Ravikovich, G.G. Vipper, V.V. Sokolovskaya et al. Publ. in B.I., 1960, N 17.
158. Pat. 3,009,879 (USA). Lubricating Grease Thickened, with Indigo and Soap Mixture. J.P. Roach. Publ. Nov. 21, 1961.
159. Kobzova R.I., Egorova Z.D., Pisarevskaya E.E. and Dmitrie-va V.G., On Efficiency of Antioxidants in Phtalocyanine Greases, Khimiya i Tekhnologiya Topliv i Masel, 1971, N 10, pp. 59-61.
160. Kobzova R.I., Egorova Z.D. and Mikheev V.A., Effect of Phtalocyanine on Lubricating Properties of Dispersion Media of Greases, Neftepererabotka i Neftekhimiya, 1971, N 11, pp. 17-19.
161. Sinitsyn V.V., Plastic Lubricating Greases in the USSR. Assortment (Refe-rence Book), Khimiya, Moscow, 1979.
162. Messina J., Rust-Inhibited Nonreactive Perfluorinated Polymer Greases, In: ASLE Proc. Intern. Conf. Solid Lubricant, Denver, Colorado, 1971. Park Ridge, 1971, pp. 326-337.
163. Messina J., Rust-Inhibited Nonreactive Perfluorinated Polymer Greases, U.S. Dep. Commer. Nat. Bur. Stand. Spec. Publ., 1976, N 452, pp. 106-118.
164. Stolarski T., Nowa koncepcija wykorzystania niskotarciowych wlasnosci policzterofluoretylenu. Zest. nauk. P. Cdan., 1975, N 243, pp. 89-101.
165. Christian J.B., Polyfluoroalkyl-Alkylpolysiloxane Grease for Instrument Lubrication, NLGI Spokesman, 1976, Vol. 39, N 11, pp. 380-385.
166. Kobzova R.N., Vyboichenko E.I. and Mikheev V.A., Antiwear and Antiseizure Properties of Polytetrafluoroethylene-Based Greases, Neftepererabotka i Neftekhimiya, 1978, N 2, pp. 16-17.
167. Panshin Yu.L., Malkevich S.G. and Dunaevskaya Ts.S., Fluoroplastics, Ehimiya, Moscow, 1978.
168. Pat. 3,112,270 (USA). Lubricants. B.Mtacek, I.P. Graham. Publ. Nov. 26, 1963.
169. Pat. 3,114,708 (USA). Dry Polyolefin/Oil Blends. A.I.Morway, &.M. Coats. Publ. Dec. 17, 1963.

170. Pat. 3,169,114 (USA). Lubricants. B.Mitacek, I.P.Graham. Publ. Peb. 9, 1965.
171. Pat. 1,351,318 (Great Britain). Grease Composition. S. Ch. Dod.son, R.H. Newman. Publ. Apr. 24, 1974.
172. Pat. 7536443 (Japan). Plastic Lubricating Grease Composition. Opubo Kapushi, Tida Dzyun, Sutyata Haltero, Nikanishi Yukio. Publ. Feb. 25, 1975.
173. Nikolaev A.F., Synthetic Polymers and Plastics Based on Them, Khimiya, Moscow-Leningrad, 1964.
174. Danishevsky S.L. and Kochanov N.N., On Toxicology of Some Organofluorine Compounds, Gigiena Truda i Prof. Zabolevaniya, 1961, N 3, pp. 3-8.
175. Danilov A.M. and Sergeeva A.V., Ureate Plastic Lubricating Greases (Collection), TsNIITEneftekhim, Moscow, 1982.
176. Pat. 4,026,890 (USA). Triasine-Urea Grease Thickeners. T.F.Wulfers. Publ. May 31, 1977.
177. Pat. 4,113,640 (USA). Triasine-Urea Grease Thickeners. T.F.Wulfers. Publ. Sept. 12, 1978.
178. Pat. 53-9243 (Japan). Polyurea Thickener-BasedLubricant. Nishida Dzonro. Publ. Apr. 4, 1978.
179. Pat. 3,879,305 (USA). Grease Thickened with Oxygen-Linked or Sulfur-Linked Polyureas. M.Ehrlich. Publ. Apr. 22, 1975.
180. Dreher J.L. and Carter C.F., New Polyurea Greases, NLGI Spokesman, 1970, Vol. 33, N 11, PP. 390-393.
181. Pat. 3,243,372 (USA). Greases Thickened with Polyarea. J.I. Dreher, J.E. Goodrich. Publ. March 29, 1966.
182. Pat. 2,121,078 (FRG). Verfahren zur Herstellung von Schmierfetten. K. Delfs, W. Moll. Publ. Nov. 16, 1972.
183. Nikishina Z.M., Strashenko M.Ya., Lendyel I.V. et al., High-Temperature Plastic Lubricating Greases Based on Urea Derivatives, In: Plastic Greases, Abstr. of Reports at 3rd All-USSR Sci. and Techn. Conf. (Berdyansk, Sept., 1979), Naukova Dumka, Kiev, 1979, pp. 227-228.
184. Pat. 4,165,329 (USA). Grease Thickening Agents. J.I. Dreher, G.M.Stanton. Publ. Aug. 21, 1979.
185. Polishuk A.T., Properties and Performance Characteristics of Some Modern Lubricating Greases, Lubr. Eng., 1977, Vol. 33, N 3, pp. 133-158.
186. Pat. 3,766,070 (USA). Diurethane-Diurea-Thickened Grease Compositions. Th. F. Wulfers. Publ. Oct. 16, 1973.
187. Pat. 3,766,071 (USA). Diurethane-Diurea-Thickened Grease Compositions. Th. P. Wulfers. Publ. Oct. 16, 1973.
188. Patzau S., Przemyslowe srodki smarove. Stan i perspektywy rozwoju, Nafta, 1981, Vol. 37, N 2, pp. 57-62.
189. Zajezierska A., Steinmiec .P. and Patzau S., Smary polimocz-nikowe, Ibid., N 9, pp. 280-285.
190. Dreher J.L. and Crocker R.E., Polyurea-Acelate Greases, NLGI Spokesman, 1976, Vol. 39, N 12, pp. 410-415.
191. Pat. 3,846,314 (USA). Grease Thickened with Ureido Compound and Alkaline Earth Metal Aliphatic Carboxylate. J.J. Dreher, C.G. Garter. Publ. Nov. 3, 1974.
192. Pat. 3,376,223 (USA). Urea Containing Grease Compositions. W. Griddle. Publ. Apr. 2, 1968.
193. Inventor's Certificate 612,951 (USSR). Plastic Lubricating Greases. P.N. Zaslavsky, Yu. S. Zaslavsky, A.M. Danilov et al. Publ. in B.I., 1978, N 24.

194. Gisser H., Petronio M. and Shapiro A., Graphite IPluoride as a Solid Lubricant, Lubr. Eng., 1972, Vol. 28, N 5, pp. 161-164.
195. Meshri D.T., Fluorographite TM Polymers, Polym. News, 1980, N 5, pp. 200-207.
196. Umanskaya O.I., Abadzheva R.N. and Yaniv A.A., Pluorinated Carbon - Thickener of High-Temperature Plastic Lubricating Greases, Neftepererabotka i Neftekhimiya, Kiev, 1985, Issue 29, pp. 56-59.
197. Pat. 2,984,624 (USA). Lubricating Oil Thickened to a Grease with a Mixture of a 1,3,5-Triazine Compound and an Organophilic Siliceous Compound. R.E. Halter, J.J. MacGrath. Publ. May 16, 1961.
198. Kobzova R.I. and Chepurova M.B., Ammeline: New Thermally Stable Thickener for Plastic Lubricating Greases, In: Plastic Lubricating Greases, TsNIITEneftekhim, Moscow, 1982, pp. 8-18 (Proc. of VNIINP, Issue 43).
199. Chepurova M.B., Kobzova R.I. and Skryabin V.P., Efficiency of Ammeline Plastic Lubricating Greases at Rolling Friction, Khimiya i Tekhnologiya Topliv i Masel, 1981, N 5, pp. 23-24.
200. Chepurova M.B., Kobzova R.I. and Mikheev V.A., Antiwear and Antiseizure Properties of Ammeline Greases, Ibid., N 9, pp. 29-30.
201. Makeeva E.D. and Makhnenko G.Kh., Sodium Amideterephtalate-Based Plastic Lubricating Greases, In: Plastic Lubricating Greases and Solid Lubricants, Khimiya, Moscow, 1969, pp. 138-145 (Proc. of VNIINP, Issue 11).

CHAPTER 4

Additives and Fillers

One of the most important techniques of upgrading the performance characteristics of lubricants, including lubricating greases, is the incorporation of various additions—additives and fillers - into their compositions [1-22]. The additions are considered as the third component of greases. The main and fundamental difference of additives from fillers is their solubility in petroleum and other oils, dispersion media of greases. Additives are surfactants, which predetermine their activity both in the bulk of grease and at the thickener-dispersion medium interface. They markedly affect the process of formation of the skeleton of soap greases and are not present in the dispersed phase. Fillers, on the contrary, should be regarded as a component of the dispersed phase of greases; they practically do not affect the structure formation.

Additives and or fillers are added to the formulation of grease not only to attain improvement in ones or the other characteristic of its quality but also to ensure an everlasting stable colloidal structure. This not only depends on the type and concentration of the addition, but is also determined by production process conditions. An attractive trend is a joint addition of additives and fillers to the composition of greases.

4.1. ADDITIVES

The solubility in oils and surface activity of additives under certain conditions significantly affect the process of structurization and the rheological characteristics of greases. It is this fact that complicates the use of additives in greases as compared to their use in oils. Nevertheless, the same additives as for the doping of oils, especially gear and transmission oils, are chiefly used also to improve the properties of greases. These include the following main groups: oxidation inhibitors (antioxidants); antiseizure; antiwear and antifriction; corrosion inhibitors and protective, viscous, thickening and adhesive additives. In addition to these main groups, also deactivators, which neutralize the catalytic action of metals on the oxidative characteristics of greases, and structure modifiers, which control the structurization process, are used. Antifoam additives, most often polysiloxanes, which enhance the efficiency of a processing vessel and prevent ejections of products from cooking kettles during removal of water from the reaction mixture are employed sometimes in production of greases.

Many additives perform several functions: thus, some antioxidants can also serve as an antiwear additive, or an additive which improves anticorrosive properties of a grease upgrades its oxidation stability, etc. Some additives, while upgrading some performance characteristics of greases, impair their other characteristics.

In a number of cases a grease contains additives of various functional actions: antioxidation and anticorrosive, antiwear and antiseizure, viscous, etc. This applies first of all to greases belonging to working-and-preservation group. A joint application of various additives may give rise to synergy; sometimes, to an insignificant relative reduction (suppression) of their functional efficiency; and sometimes, to antagonism; this should be taken into account while selecting the composition of additives. This possibly encouraged the development and employment of special additives, so-called additive packages, for improving the performance characteristics of, e.g., transmission oils. The packages are sometimes used for the doping of greases.

The widest use has been found by antioxidants not in the amount of their consumption, but in the number of greases containing such additives. This is accounted for, firstly, by the fact that nearly all soaps are oxidation catalysts [12,23,24] and hence the greases must be inhibited; secondly, severe temperature conditions of application of many modern greases, accelerate their oxidation; and, thirdly, striving to extend the duration of storage and operation life of greases and to upgrade the economic efficiency of their production and application. Modern greases should not be produced without antioxidation additives.

Most antioxidation additives, exhibiting an adequate efficiency at concentrations of 0.3-1 %, exert no significant effect on the structure and rheological properties of greases. When choosing the grease production process stage where it is expedient to add the antioxidants, their thermal stability should be the deciding factor; it is more effective to add the additives at the cooling stage [20].

Various compounds have been proposed as antioxidants for greases [6-13,15-33]: diphenylamine, thiodiphenylamine, phenyl-a-naphthylamine, phenyl-b-naphthylamine, paraoxydiphenylamine, naphthylamine, aldol-a-naphthylamine, b-naphthol, 2,6-di-tertiary-butyl-4-methyl-phenol, 4,4'-methylene-bis-2,6-di-tertiary-butylphenol, 2,2'-methylene-bis-4-methyl-6-tertiary-butylphenol, tertiary aminoamides of isostearic acid, dimethylethylenebenzylphenylureate, dode-cylbenzylureate, alkylphenylphosphate, trisodium phosphate, dilauryl selenide, dithiocarbamates of metals and many other compounds. Characteristics of the substances used at present for inhibition of greases are presented in Table 4.1 [13,15-19,24,33].

A large group of additives used to improve the performance characteristics of greases operating at high specific loads is formed by antiseizure (or extreme pressure) and antiwear additives [2-5, 9-13, 19-21,33-39]. The former prevent scoring, alleviate seizures and reduce the destruction of metal surfaces, reducing the friction at the same time, while the latter prevent a progressive wear of metals and reduce the friction, some of them stabilizing the latter by eliminating an irregular character of the friction process, an intermittent sliding. Additives reducing and stabilizing the friction are also called antifriction additives. On the whole, the group of antiseizure, antiwear and antifriction additives enhances the load carrying capacity of the lubricating film of greases and their lubricity.

It should be pointed out that the lubricity of most greases is higher than that of the oils with which they have been prepared, since some thickeners (soaps, especially complex ones, as well as pigments, carbon black, oleophilic graphite, polymers, etc.) serve as antiseizure, antiwear, and antifriction additives. Because of this, chemically active antiseizure and antiwear additives should be used only in exceptional cases for greases used in heavily loaded lubrication points.

ADDITIVES AND FILLERS

Table 4.1. Antioxidation additives

Additive	Formula	Melting point, °C	t_1, °C	$\dfrac{t_{10\%}}{t_{50\%}}$	$\dfrac{M_{150}}{M_{200}}$	i	Remark
Fenam (diphenylamine) GOST 194-80	H–N(phenyl)(phenyl)	58-62	195	148/183	12/84	11-12	Weakly toxic (MPC=10 mg/m³); flash point ≥165°C. At concentration of 0.5-1.5% does not break down grease structure.
DAT (diphenylamine alkylated with propylene trimers) TU 38 401175-82	H–N(Ar-R)(Ar-R)	—	250	230/280	0/2.5	15-20	Slightly toxic. At concentration of 0.5-1.5% does not break down grease structure
Fentiazin (thiodiphenylamine) TU 6-14-322-76	phenothiazine	175-180	240	198/245	1/11	85-90	Less-toxic. At concentration of 0.5-1.5% does not break down grease structure
Neozon-A; Naftam-1 (phenylnaphthylamine) TU 6-14-202-74	H–N(phenyl)(1-naphthyl)	55-60	257	203/247	1.7/9.3	70-75	Weakly toxic (MPC=3 mg/m³). At concentration of 0.5-1.5% does not break down grease structure
Neozon-D; Naftam-2 (phenylnaphthylamine) GOST 39-79	H–N(phenyl)(2-naphthyl)	105-108	260	213/255	0.5/5	70-75	Weakly toxic (MPC=1mg/m³); flash point 205°C; ignition point 238°C. At concentration of 0.5-1.5% does not break down grease structure
Naphthylamine GOST 8827-74	1-naphthyl-NH₂	50	190	150/180	12/77	10-15	Highly toxic as compared with other antioxidants. At concentration of 0.5-1.5% does not break down grease structure

Additive	Formula	Melting point, °C	t_1, °C	$\dfrac{t_{10\%}}{t_{50\%}}$	$\dfrac{M_{150}}{M_{200}}$	i	Remark
Aldol- naphylamine GOST 830-75	N=CH—CH$_2$—CH—CH$_3$ / OH, naphthyl	60-70	205	166/187	9/67	35-40	as above
OM-DETA (product of condensation of commercial alkylphenols, ethylene tria-mine with formaldehide) STP 01.11.09-4.01.145-84	[OH, (CH$_2$)$_2$—N—(CH$_2$)$_2$, R]$_2$ NH	—	243	198/265	3/11	32-35	Less-toxic. 50-% solution in oil with flash point 170°C. At concentration of 0.5-1.5% does not break down grease structure
Borin (mixture of bis- and tris-, mono- and dialkyl-4-oxybenzyl- amines) TU 38-1011003-84	C—(CH$_3$)$_3$ H / HO—CH$_2$—N—CH$_2$—OH	—	209-263	185/284	5/13	35-40	Less-toxic. 50-% solution in oil with flash point 160°C.
Naphyol (2-naphtol) GOST 5835-79	naphthyl—OH	115-120	173	120/150	9/76	5-7	Non toxic
Ionol (2,6-di-tert-butyl-4-methylphe-nol) GOST 10894-76	(CH$_3$)$_3$C—[OH, C(CH$_3$)$_3$]—CH$_3$	40-60	155	112-144	65-94	7-8	Non toxic

Additive	Formula	Melting point, °C	t_1, °C	$\frac{t_{10\%}}{t_{50\%}}$	$\frac{M_{150}}{M_{200}}$	i	Remark
LZ-MB-1 (4,4'-methylene-bis-2,6-di-tert-butyl-phenol) TU 38-101441-74	(structure)	150-155	243	225/260	0.5/3.5	45-53	**Less-toxic** (MPC = 5 mg/m³); flash point 220°C; ignition point 235°C. At concentration of 0.5-1.5% does not break down grease structure.
Agidol-2 (2,2'-methylene-bis-4-methyl-6-tert-butyl phenol) TU 38-101617-80	(structure)	125-130	235	215/254	0.5/5	35-40	Non toxic
Molyvan A (molybdenum dithiocarbamate)	$\left[\begin{array}{c}R\\R\end{array}\!\!N-\overset{S}{\underset{}{C}}-S\right]MO_2O_x\,Sg$	251-255	—	—	—	> 100	Non toxic

Additive	Formula	Melting point, °C	t_1, °C	$\dfrac{t_{10\%}}{t_{50\%}}$	$\dfrac{M_{150}}{M_{200}}$	i	Remark
Vanlube AZ-B (zinc diamyldithiocarbamate)	$\left[\begin{array}{c} C_5H_{11} \\ \\ C_5H_{11} \end{array} N - C \begin{array}{c} S \\ \parallel \\ \end{array} - S \right]_2 Zn$	—	—	—	—	> 100	Solution of additive oil with flash point 160°C
Etiltsimat (zinc diethyldithiocarbamate) TU 6-14-809-77	$\left[\begin{array}{c} C_2H_5 \\ \\ C_2H_5 \end{array} N - C \begin{array}{c} S \\ \parallel \\ \end{array} - S \right]_2 Zn$	—	—	—	—	35–40	Highly toxic as compared with other antioxidants
Vanlube-674		—	—	—	—	> 100	Solution of additive in oil with flash point 174°C

Note. t_1—additive evaporation and decomposition temperature by DTA, °C; $t_{10\%}$ and $t_{50\%}$—temperatures of loss of 10 and 50% of the additive mass respectively by DTA, °C; M_{150} and M_{200} °C respectively, %; i—induction period of oxidation at 120°C of petroleum oil-based lithium grease containing 1% of the additive, hours.

Used as antiwear and antiseizure additives for greases can be organic compounds and other products, containing active atoms of sulfur, chlorine, phosphorus, such as sulfurated natural fats and petroleum-origin products, di- and polysulfides, xanthates, di-thiocarbamates, chloro derivatives of various organic compounds and other products, dialkyldithiophosphates of metals, di-(alkylphenyl)-dithiophosphates of metals, triphenyl and tricresi phosphates, thiourea derivatives, etc.,as well as such products like salts of molybdic and tungstic acids, alkyl selenides, lead napththenates, cadmium salts of acetic or oxalic acid, etc. [4, 9-13,15,17, 19-21, 33-39]. Characteristics of some antiseizure and antiwear additives used in greases are presented in Table 4.2 [4, 10, 13, 15, 17, 19, 33, 35, 36, 38, 39].

Additives containing sulfur, chlorine and phosphorus are chemically active and often cause negative side effects, due to which they, with the exception of sulfurated natural (vegetable and animal) fats and petroleum-origin products, have found no wide use. These additives are opposed by such competing additions-fillers as graphite, molybdenum disulfide, featuring a high lubricating effect while not impairing significantly other grease characteristics.

It is expedient to add antiseizure and antiwear additives to greases in amounts of 2-5% at the cooling or homogenisation stage. It should be taken into account that the efficiency of additives not only depends on their composition and properties, but is also greatly determined by the qualitative and quantitative compositions of the grease. At the same concentrations of additives, their effect is higher in nonsoap greases.

Anticorrosive and protective additives [6-13, 15-20, 40-47] are added to the composition of greases to improve their protective properties and to prevent corrosion of metallic products and friction units. Anticorrosive additives prevent chemical action of corrosive components of a grease on metals, while protective additives prevent the corrosion of metals under the action of a corrosive environment, i.e., protect the metal from an electro-chemical corrosion in presence of an electrolyte. These additives are used to improve the properties of preservation and working-and-preservation antifriction greases. Their selection depends on many factors, of which the grease composition, nature and properties of metals to be protected, and operating conditions (temperature, environment, etc.) are most important. Anticorrosive additives often play the part of protective ones, and vice versa. The following main types of surface-active products and compounds are used as anticorrosive and protective additives in greases: oxidized petrolatum and ceresin; nitrated oxidized petrolatum; amines, amides, imides, sulfonates, and alkylsalicylates of calcium and barium; salts of naphthenic and sulfonic acids; complex compounds of amines and S.F.A; urea alkylsuccinimides; derivatives of benzotriazole and of mercaptobenzotriazole; dextramin; lanolin; dialkyldithio-phosphates; sodium nitrite; sulfides and disulfides, etc. Characteristics of the compounds used in production of protective and working-and-preservation greases are presented in Table 4.3 [6-8, 15,15-19,40,46].

Some corrosion inhibitors and protective additives, while performing their main function, improve at the same time the anti-oxidation properties and lubricity of greases, whereas other ones exert adverse side effects: weaken the system. These additives are added to greases (1-5 %) at different production process stages, depending on the grease composition and additive type.

When improving the adhesive characteristic of greases, it is possible also to upgrade their protective properties. This is attained by adding to greases various polymeric additions,

Table 4.2. Antiseziure and antiwear additives

Additive	Formula	P_{cr}, N	P_w, N	Wear index	Remarks
1	2	3	4	5	6
OkhM (sulfurated cottonseed oil), antiseizure, TU 38.30171-74	$\begin{array}{c}\text{RCH——CHR'}\\ \diagdown\!\!\diagup\\ S\end{array}$; $\begin{array}{c}\text{RCH——CHR'}\\ |\;\;\;\;\;\;\;\;\;\;|\\ S\;\;\;\;\;\;\;\;\;\;S\\ |\;\;\;\;\;\;\;\;\;\;|\\ \text{R''CH——CHR''}\end{array}$	Calcium grease + 5% additive 800-900	1200-1500	30-35	Nontoxic; does not break down structure; less corrosive; content: S = 6-10%;
OKZh (sulfarated sperm whale fat, antiseizure	as above	Lithium grease + 3% additive 700-800	1800-2100	30-75	Nontoxic; does not break down structure; less corrosive; content: S = 9-11%;
KINKh-2 (alkylpoly-sulfide), antiseizure, TU 38.101980-84	$\left[\begin{array}{c}(CH_3)_2-C-CH_2-\\ |\\ S_2\\ |\\ -CH_2-C-(CH_3)_2\end{array}\right]_n$ $n = \geq 3$	Lithium grease + 5% additive 1200-1300	2500-3500	75-80	Does not break down structure; t = 236-238°C; t_{10} = 138°C; t_{50} = 224°C; M_{150} = 13%; M_{200} = 29%; MPS = 9 mg/m³; content: S = 40 − 45%; slightly corrosive;
LZ-23K (ethylene diisopropy-xanthate), antiseizure, GOST 11883-77	$\begin{array}{c}\;\;\;\;\;\;\;\;\;S\\ \;\;\;\;\;\;\;\;\;\|\\ CH_2-S-C-O-C_3H_7\text{-iso}\\ |\\ CH_2-S-C-O-C_3H_7\text{-iso}\\ \;\;\;\;\;\;\;\;\;\|\\ \;\;\;\;\;\;\;\;\;S\end{array}$	Lithium grease + 3% additive 800-900	2600-2900	40-45	Slightly breaks down structure; strongly corrosive in presence of moisture; content: S = 38-43%;
Sulfol (di-(5,5,5-trichloroamyl)–sulfide, antiseizure, MRTU 6-03-113-64	$[CCl_3-(CH_2)_2]_2S$	Lithium grease + 3% additive 1000-1200	2000-2500	50-60	Does not break down structure; t = 230-235°C; t_{10} = 145°C; t_{50} = 220°C; M_{150} = 11%; M_{200} = 28%; content: S = 7-9%, Cl =

ADDITIVES AND FILLERS

Additive	Formula	P_{cr}, N	P_w, N	Wear index	Remarks
1	2	3	4	5	6
52-57% KhSO-200 (b,b-dichlordialkyl-disulfide), antiseizure, TU 38.101828-80	R—CHCl—CH$_2$ | S$_2$ | R—CHCl—CH$_2$	Lithium grease + 5% additive			Corrosive; weakly toxic; $t = 224$-$234°C$; $t_{10} = 187°C$; $t_{50} = 224°C$; $M_{150} = 3\%$; $M_{200} = 20\%$; MPS = 5 mg/m^3; content: S = 14-16%, Cl = 14-16%
		900-1200	1800-2000	46-48	
KhP-470 (chlorinated paraf-fin), antiseizure, TU 6-01-568-76	—	Lithium grease + 5% additive			Improves also antiwear properties; content: Cl=40-50%;
		1000-1100	2500-3000	55-60	
Sovol (mixture of tetra- and pentachloro diphenyl), anti-seizure, OST 6-01-24-75	—	Lithium grease + 5% additive			Content: Cl = 40-45%
		850-900	2000-2500	40-42	
Khloref-40 (o,o-dibutyl ether of trichloro-methylphosphinic acid), antiseizure TU 6-02-579-75	CCl_3—$\overset{\overset{O}{\|}}{P}$—$(OO_4C_9)_2$	Lithium grease + 5% additive			less-toxic; weakens structure of cCa-greases; d_4^{20} = 1.16-1.22; content: Cl = 28-34%, P=10%;
		900-1000	1800-2000	46-48	
LZ-318 (trichloroamyl ether of 2-mercapto-benzothriazole), antiseizure, TU 38.101541-75	benzothiazole—S(CH$_2$)$_4$—CCl$_3$	Lithium grease + 5% additive			Does not break down structure; $t_{10} = 95°C$; $t_{50} = 233°C$; $M_{150} = 31\%$; $M_{200} = 37\%$; content: S = 15-20%, Cl-20-25%; others, 4-5%
		850-950	1200-1500	36-38	

Additive	Formula	P_{cr}, N	P_w, N	Wear index	Remarks
1	2	3	4	5	6
LZ-309/2 (product of inter-action of sodium diisopropyl-dithiophospate with tetraclor-pentane), antiseizure, anti-wear, TU 38.101748-78	$\begin{array}{c}\text{CH}_3\text{—CH—O}\quad\text{S}\\ \quad\quad\quad\text{CH}_3\quad\diagdown\!\!\diagup\\ \quad\quad\quad\quad\quad\text{P}\\ \text{CH}_3\text{—CH—O}\quad\diagup\!\!\diagdown\text{S—(CH}_2)_4\text{CCl}_3\\ \quad\quad\quad\text{CH}_3\end{array}$	Lithium grease + 5% additive 1000-1100	1500-1800	42-45	Does not break down structure; $t = 191\text{-}193°C$; $t_{10} = 140°C$; $t_{50} = 191°C$; $M_{150} = 12\%$; $M_{200} = 67\%$; content: $S = 12.8\%$, Cl- 23.0%; others, 6.8%
Tricresyl phosphate, antiwear, GOST 5728-76	$(CH_3\text{-}C_6H_4O)_3PO$ (mixture of o-, m-, n-isomers)	Lithium grease + 2% additive 600-700	1300-1600	28-30	Toxic, especially ortho-isomer; $d_4^{20} = 1.117$; $t = 281\text{-}285°C$; $t_{10} = 233°C$; $t_{50} = 273°C$; $M_{150} = 1\%$; $M_{200} = 3\%$; content: $P = 84\%$
KASP-13 (barium salts of derivatives of arylalkyl-dithio-phosphoric acids), antiwear, antiseizure, TU 38.101831-85	[barium arylalkyldithiophosphate structure, Ba]$_2$	Complex lithium grease + 2% additive 850-950	2800-3500	40-52	Improves anticorrosive and anti-oxydising characteristics. Does not break down structure. Content: $S = 1.2\text{-}1.8\%$; $P = 0.6\text{-}1.6\%$
VNIINP-354 (di(alkylphenyl)-dithiophosphate of zinc), anti-seizure, antiwear, antioxidation, TU 38.101680-82	[zinc di(alkylphenyl)dithiophosphate structure, Zn]$_2$	Lithium grease + 5% additive 850-950	1100-1700	36-40	$t_i = 14\text{-}15h$; content: $S = 7\text{-}8\%$, $P = 4\text{-}5\%$; others, 4-5%; thermally stable up to 280°C; weakly breaks down structure

ADDITIVES AND FILLERS

Additive	Formula	P_{cr}, N	P_w, N	Wear index	Remarks
1	2	3	4	5	6
DF-11 (zinc dialkyl dithiophosphate), antiwear, antiseizure, antioxidation, GOST 24216-80	$\left[\begin{array}{c} RO \\ RO \end{array} P \begin{array}{c} S \\ S- \end{array}\right]_2 Zn$	Lithium grease + 5% additive			Weakly breaks down structure: t_i = 35-40h; t = 238-242°C; t_{10} = 205°C; t_{50} = 226°C; M_{150} = 2%; M_{200} = 8%; slightly corrosive
		800-900	1100-1700	36-40	
VIR-1, antiseizure, antiwear, anticorrosive, TU 38.401799-83	Composition consisting of seven components containing S, P, N	Lithium grease + 5% additive			Does not breaks down structure; content: S = 29-31%; P = 1.5-1.8%; others, 0.8-1.0%
		1200-1300	2500-3000	60-70	
Anglamol-99, antiseizure, antiwear, anticorrosive, anti-oxidant	Composition of additives containing S, P, N	Lithium grease + 5% additive			Does not breaks down structure; content: S = 31-35%; P = 1.7-2.0%; others, 0.1-0.5%
		1200-1300	2500-3000	60-70	
Lead napthanate, antiwear, antifriction	Lead salts of naphthenic acods (acidols or distilled acids)	Lithium grease + 3% additive			Toxic; does not break down structure
		600-900	1800-2000	34-38	
Makoma-76 (2.5% lead napthanate + 2% sulfated products), antiseizure, anti-wear, synergistic, antifriction	Sulfurated products, e.g., sulfurated castor, sperm, and other oils	Lithium grease + 5% additive			as above
		900-1000	2800-4000	50-60	

Note. Designation are the same as in Table 4.1.

Table 4.3. Anticorrosive additives

Additive	Formula	Protective properties				Remark
		Chamber G-4, % corroded surface (degrees) Sea water	% corroded surface (degrees)	"Dynacorrotest" method, corrosion rate, mm/year (A/cm^2), at rotation speed ω = 1500 rpm	Multielement detector, % corroded surface. Steel 45, sea water, 24 h	
NOP (nitrated oxidized petrolatum)	Complex mixture of nitrated high-molecular isoparaffin and naphthenic acids and aromatic acids with long side isoparaffinic redicals and of other oxygen-containing compounds in petroleum oil	100 cycles 9-11 (8)	Lithium grease + 1% additive; on steel 45 7 days 38-42 (10)	0.00025 (2.5·10^{-8})	—	Viscous brown products, poorly soluble in oils; AN before neutralization: 55-80 mg KOH; Improves lubricating properties of grease
MNI-7 (oxidized ceresin), GOST 10584-79	Complex mixture of high-molecular isoparaffin and naphthenic acids and aromatic acids with long side radicals and of other oxygen-containing compounds with insignificant content of oils	9-10 (8)	Hydrocarbon grease with 3% additive; on steel 45 1-3 (6)	0.00020 (2·10^{-8}) with 2% 0.00100 (1·10^{-7})	6-10	Brown product, insoluble in oils; melting point ≥60°C; AN = 55-75 mg KOH/g; SN = 120-140 mg KOH/g; improves lubricating properties; does not weaken grease structure
KAP-25 (mixture of alkylsuccinic acids), TU 6-14-93-81	C_nH_{2n-1}—CH—COOH | CH_2—COOH N—12-15	Oil ASV-5 + 5% additive; Lithium grease +1% additive; on steel 45 24h 20h 0 2-4 (0) (6)		0.00300 (3·10^{-7})	15-20	Flash point ≥ 180°C; ignition point ≥385°C; AN = 350-385 mg KOH/g; does not weaken grease structure

ADDITIVES AND FILLERS

1	2	3	4	5	6	7
SIM (urea alkenyl-succinimide), TU 38.401111-84	$R = C-CH_2-CH$ with CH_3, $C=O$, $N-CONH_2$, $CH_2-C=O$	Lithium grease + 1% additive; on steel 45 100 cycles 2-4 (6)	Lithium grease + 1% additive; on steel 45 7 days 14-16 (9)	0.00250 $(2.5 \cdot 10^{-7})$	Lithium grease + 2% additive 20-40	Flash point $\geq 200°C$; ignition point $\geq 270°C$; AN = 6.5 mg KOH/g; Somewhat weakens grease structure; improves antiseizure and corrosion stability of greases
M-1; M-2 (complex salts of dicyclohexylamine with SFA) TU 6-02-742-73	(R for M-1: $C_{10}-C_{19}$; R for M-2: $C_{17}-C_{20}$) cyclohexyl$-NH_2-HOOC-R$	Lithium grease + 1% M-11; on steel 45 100 cycles 1-2 (6)	7 days 18-22 (9)	0.00500 $(5 \cdot 10^{-7})$	Lithium grease + 2% M-11 on steel 45 10-20	Insignificantly weaker grease structure (2-5%); improve antioxidation properties
KhT-3 (ester of alkenyl succinic acid with dimethylol urea)	$R-$phenyl$-S(=O)-O-CHNHCH_2OH$	Oil AS-6 + 5% additive; on steel 45; 200 h 6-8 (8)	20 h 4-5 (7)	Lithium grease + 2% additive; on steel 45 0.00280 $(2.8 \cdot 10^{-7})$	Lithium grease + 2% additive; on steel 45 20-35	Weaker structure and lower drop point of greases
Sulin	Sulfoacids (S-300) + calcium salt of alkylbenzenes + tricresyl phosphate + oleic acid + calcium alkyl phenolate + benzotriazole	Oil AS-6 + 5% additive; on steel 45; 200 h 0(0)	20 h 0(0)	Lithium grease + 1% additive; on steel 45; 0.00160 $(2.8 \cdot 10^{-7})$	Lithium grease + 1% additive; on steel 45 40-45	Same as above
MASK (alkaline calcium alkyl salicylate), OST 38-01100-76	Ca salicylate structure: OH, $COO-Ca-OOC$, $R = C_8-C_{12}$	—	—	Lithium grease + 1% additive; on steel 45 (0.00380) $3.8 \cdot 10^{-7}$	Lithium grease + 1% additive; on steel 45 50-60	50-60% solution in oil AS-6; improves antioxydation properties of greases; does not break down their structure; AN = 100 mg KOH/g; d_4^{20} = 1.124; melting point 21°C

1	2	3	4	5	6	7
INGA-1 (nitrared alkenyl succinimide), TU 38-4031-72	$\begin{array}{c}\text{CH}_3\quad\text{O}\\R\text{-CH-C-CH}_2\text{-CH-C}\\\quad\;\;\mid\quad\;\mid\quad\;\;\mid\quad\;\;\;\;\;\;\text{N-CONH}_2\\\text{HO NO}_2\quad\text{CH}_2\text{-C}\\\quad\quad\quad\quad\quad\quad\;\;\text{O}\end{array}$	Luthium grease + 5% additive; on steel 45 — 100 cycles 4 – 6 (7)	Luthium grease + 5% additive; on steel 45 — 7 bays 24-28 (9)	—	—	AN = 15 mg KOH/g; viscosity, 120 mm²/s at 100°C; somewhat weakens grease structure
Triethanolamine, TU 6-02-916-79	$(HO\text{-CH}_2\text{-CH}_2)_3\text{-N}$	Luthium grease + 1% additive; on steel 45 — 100 cycles 2 – 3 (6)	Luthium grease + 1% additive; on steel 45 — 7 bays 12-15 (9)	0.00320 (3.2 . 10⁻⁷)	—	Colorless viscout liquid with weak ammonium odor; boiling point, 360°C; melting point, 21.2°C; $d_4^{20}=1.124$
TAL-3 (product of condensation of tall oil fatty acids, deithylenetriamine with paraform and alkylphenol), TU 38-30-241-76	(structure: CH₂-CH₂-NH-CH₂ group with phenol OH-R₁, and CH₂-N=CR, CH₂-N=)	—	—	Lithium grease + 2% additive; on steel 45 0.00360 (3.6 . 10⁻⁷)	—	Somewhat weakens structure of greases, improves their antioxydation properties
DEKSTRAMIN, TU 64-6-184-81	(structure: OH-C₆H₃(NO₂)-CH-CH-CH₂-OH with NH₂)	—	—	Lithium grease + 2% additive; on steel 45; 0.00420 (4.2 . 10⁻⁷)	—	Melting point, 155-160°C (dextrorotatory)
LANOLINE		Oil ASV-5 + 5% additive; on steel 45 24 h 0(0)	20 h 9-11(8)	—	—	Wool fat

such as latex, rubbers, vinypol, octol, atapol, polyisobutylene, polymethacrylates and other polymeric additions [9,11, 15-17, 19,20,48]. The main function of oil-soluble polymeric additions is, however, upgrading the viscous, adhesive and protective properties of greases; they significantly improve the antifriction properties of greases and in some cases, at higher concentrations, serve as the thickener in polymer greases, and can therefore be classified as semifunctional additives. Some characteristics of the viscous and adhesive additives used to improve the grease properties are given in Table 4.4 [13,16,17,48,49]. An advantage of these additives is that they do not reduce the thickening effect of the main grease thickener and do not impair the colloidal and rheological characteristics of greases. They are added at the beginning of the grease preparation process together with charging the first amount of the dispersion medium. Viscous and adhesive additives are sometimes dissolved beforehand in the dispersion medium.

We classify solid high-polymer additions, poorly soluble or fully insoluble in oils and containing chemically bound halogens, such as polytetrafluoroethylene, etc., with fillers. Some information on polymeric additions as grease thickeners is presented in Chapter 3.

Small amounts of special additives called structure modifiers, can be added to control the grease formation process in production of greases [6,7,12,19,20, 50,51]. These are free fatty acids, high molecular weight mono- and polyhydric alcohols, ethers, naphthenates of metals and salts of other organic acids, oil-soluble sulfonates, amines, alkenylsuccinimides and other surfactants.

Water, being a structure stabilizer in hydrated Ca-greases, as also in other soap greases where it is in insignificant amounts and is coordination-bound to the thickener, serves as a structure modifier as well. Many of the above-listed modifiers (fatty acids, alcohols, glycerol, alkalis) are unavoidably present in soap greases as they are formed in their production process. Moreover, all petroleum oil-based greases contain resins as well as products of oxidation of dispersion media, formed chiefly at the thermal treatment stage.

The structure modifiers, being surfactants, feature a high- energy bond with main grease components, particularly with thickeners. They exhibit surface active properties both at the thickener dispersion medium phase boundary and in the bulk of the system.

The action of structure modifiers in soap greases consists, firstly, in their peptizing properties and participation in the formation of primary elements of the colloidal system (structure), namely associates and micelles, with the result that the critical association and micelle concentrations, solubilizing ability of the soap, its polarity and polarizability, etc. change, which in turn changes colloidal characteristics of the soap thickener. Secondly, the modifiers, at sufficient concentration, being adsorbed on arising crystallization centres, on elements of the skeleton (this occurs at all stages of its formation), form adsorption solvate layers on the thickener surface, which affects the total balance of energetic bonds in the system (grease), mainly reducing it.

The addition of small amounts of a modifier to soap greases increases the dispersity and enhances the thickening effect of the thickener, which increases the strength and upgrades the colloidal stability of the grease, while considerable concentrations of the modifier impair these characteristics.

Table 4.4. Viscous and adhesive additives

Additive	Formula	Molecular weight (M)	Function		Remark
			Main	Accompanying	
1	2	3	4	5	6
Octol-600 (product of polymerization of isobutylene and N-butenes), TU 38.101339-73	—	~600	Viscous	Adhesive	Viscous neutral liquid of yellow to light brown color; viscosity at 100°C, 550-650 mm²/s; flash point ≥220°C
Vinipol VB-2 (VB-3) (product of polymerization of vinyl-n-butyl enter), TU 6-01-744-77	$\left[\begin{array}{c} -CH_2-OH- \\ \| \\ OC_4H_7 \end{array}\right]_n$	4000-12000	Viscous	Adhesive	Viscous transparent liquid of light-yellow to light brown color; concentrates (TU 6-01-1248-80) are also produced, which are mixtures of VB-2 (40-60%) or VB-3 (50-70%) in petroleum oil and have flash point ≥92°C and congelation point ≤ 30°C
Polyisobutylene P-20. TU 38.103257-80	$\left[\begin{array}{c} CH_3 \\ \| \\ -CH_2-C- \\ \| \\ CH_3 \end{array}\right]_n$	15000-25000	Viscous	Adhesive	Viscous thick mass; concentrates KAP-10 (M=9.000-15.000; concentration, 30-35%) and KAP-20 (M=15.000-20.000; concentration, 25-30%) are also produced; liable to mechanical destruction, whose degree is determined by viscosity of thickened oil and decreases with increasing viscosity of oil

ADDITIVES AND FILLERS

Table 4.4. (continued)

1	2	3	4	5	6
Low-pressure high-density polythylene, GOST 16338-77	$(-CH_2-CH_2-CH_2)n$	—	Viscous; when added to grease as powder, serves as filler	Adhesive; thickening; antifriction	Dence white mass (d_4^{20} =0.945 – 0.954; melting point, 132°C); dissolving in petroleum oils at 125-135°C; at 140°C decomposes with evolution of volative organic acids, form- and acetal-dehydes, and carbon oxide; in powdered state forms exposive mixtures with air, whose lower limit of explosivity is of 18-45 g/m^3
Atapol (atactic polypropylene)	$\left[-CH_2-CH_2-\ \vert\ CH_3\right]_n$	—	Viscous; when added to grease as powder, serves as filler	Thickening; antifriction	—
Polybutene	$\left[-CH_2-CH-\ \vert\ C_2H_5\right]_n$ $n = 200\text{-}400$	—	Viscous	Thickening	—

Table 4.4. (continued)

1	2	3	4	5	6
Polymethacrylates (V and D)	$\left[\begin{array}{c} CH_3 \\ | \\ -CH_2-C- \\ | \\ C=O \\ | \\ OR \end{array} \right]_n$ $n = 30-80$ R-aliphatic radical $C_{12}-C_{18}$	15000	Viscous	Grade "D" has also depressor properties	Produced as 30-40% solutions of additive in petroleum oil 1-20
Synthetic rubber: SCD (polymerized butadiene); SKI-3 (polymerized izoprene); SCS-30 (copoly-merization of diene and vinyl compounds: butadie-nestyrene rubber, ets.)	—	SKD: $(0.7-2.3) \cdot 10^5$; SCI-3: $(3.5-13) \cdot 10^5$; SCS-30 $(2.0-2.2) \cdot 10^5$	Viscous	Adhesive, protective	Rubbery substances fearing high elasticity over temperature range from −70 to 120°C and dissolving in oils
Colophony, GOST 19113-73 or GOST 14201-83	Mixture of resin (abietic) unsaturated acids (90-95%) and other fatty acids (5-10%) of general formula $C_{19}H_{29}COOH$	—	Adhesive	Anticorrosive protective	For detailed characteristics, refer to Chap.3, Table 3.16

ADDITIVES AND FILLERS

The action of structure modifiers in nonsoap greases shows up in their ability to be adsorbed on an extensive surface of organic and inorganic thickeners and, naturally, to affect the energy of bonds in the system.

Having confined the discussion to the above, let us note that most antioxidation, antiseizure, antiwear and antifriction, anti-corrosive, protective, and other additives are surfactants. They should therefore be regarded as structure modifiers, and results of the doping of oils with additives should not be extended to lubricating greases. The introduction of additives in the grease preparation process should not be arbitrary; optimum conditions should be determined for every case, taking into account the composition and properties of the dispersion medium, dispersed phase, additive itself or their combination, etc.

It would be wrong to believe that the use of additives of various functional purposes, especially of their "packages", is necessary and useful in grease production. One should remember that many thickeners significantly improve the antiseizure, antiwear, antifriction, protective and other properties of the lubricating liquids-oils, thickened by them. Thus, all soaps, particularly complex ones, as well as high polymers improve the lubricity of oils; urea derivatives enhance the protective and antioxidation properties; many pigments serve as good antioxidants and additionally perform the function of antiwear additives, etc.

A detailed information on additives for greases, their properties and efficiency of action is presented in [1-9, 15, 19, 20, 40, 42]. It should be noted, however, that, despite numerous publications dealing with the use of additives in greases, these problems have not been studied in full; different literature sources contain sometimes contradictory data as well as conclusions not confirmed by experiments and practice.

4.2. FILLERS

Fillers, or solid additions, are highly dispersed oil-insoluble materials which do not form a colloidal structure in greases, but are an independent dispersed phase and improve the performance characteristics of greases [9-13, 19, 52-69]. High lubricity, chemical and thermal stabilities are the main causes of an extensive use of fillers as additions to greases operating in severely loaded lubrication points and under other unfavourable conditions.

The widest use has been found of crystalline layered fillers featuring low friction coefficients: molybdenum disulfide and diselenide, graphite, mica, talc, boron nitride, vermiculite, sulfides of some metals and other materials. Solid high-polymer additions containing bound halogen atoms, such as polytetrafluoroethylene (PTEE), metal oxides, metallic powders are employed as well.

Molybdenum disulfide (MoS_2) is a powder of a black colour with a greyish shade. Natural MoS_2 crystallizes into a hexagonal form and has a layered structure. Synthetic MoS_2 has a layered structure as well, but either of a rhombohedral form or of a form intermediate between a hexagonal and a rhombohedral one. Molybdenum atoms are interposed between two sulfur atoms spaced from each other at 3×10^{-10} m (3A), the closest distance between molybdenum and sulfur atoms being 2.41×10^{-10} m (2.4·1 A). The basic properties of molybdenum disulfide are as follows [11, 52, 53]:

Density → kg/m³ 4800
Mohs hardness number → 1.0-1.5
Initial sublimation temperature → 450°C
Melting point, °C → 1185
Stability to oxidation in air or oxygen, → upto 300-350
Radiation stability, → stable upto 5×10^9 m
Dry friction coefficient at temperatures up to 550°C → 0.02 to 0

At a temperature of 400-600°C molybdenum disulfide oxidizes in air, forming solid abrasive MoO_3. It also oxidizes by interaction with concentrated H_2SO_4 and HNO_3, dissolves in aqua regia, vigorously reacts with fluorine and on heating, with chlorine, transforming into MoF_2 and $MoCl_2$, but does not practically react with bromine. Hydrogen reduces it to metal. Heating of MoS_2 in vacuum results in formation of Mo_2S_3.

Milling of molybdenum disulfide for 600 h in a colloidal mill does not change the parameters of its crystal lattice. It retains good lubricating properties in vacuum in absence of moisture, whose presence impairs these properties. A thin MoS_2 layer on a friction surface withstands dynamic loads up to 10^3 MPa, and under static conditions up to 3×10^3 MPa.

Table 4.5. Characteristics of molybdenum disulfide (TI 48-18-133-25)

Characteristics	MoS_2 grades			
	DM-1	*DM-2*	*DM-3*	*DM-4*
Content of molybdenum disulfide, % (not below)	99.68	99.62	99.35	99.20
Content of impurties, % (not over):				
silicon	0.02	0.05	0.05	0.10
iron	0.10	0.10	0.20	0.20
aluminium	0.03	0.03	0.10	0.20
calcium	0.07	0.10	0.20	0.20
oxidized molybdenum	0.10	0.10	0.10	0.10
Content of particles, %, with conventional diameter:				
$7 \cdot 10^{-6}$ m and less (not below)	99.0	—	99.0	—
$140 \cdot 10^{-6}$ m and more (not over)	—	0.01	—	0.01
Content of moisture and oil, % (not over)	0.50	0.50	1.00	1.00
Reaction of aqueous extract (pH)	5.4-6.6	5.5-6.6	5.4-6.6	5.4-6.6

Owing to the good lubricating properties, MoS_2 is widely used in production of lubricants. It is mainly employed to prepare solid lubricating coatings, oil suspensions and pastes and also as an antifriction addition to lubricating greases. Specially processed MoS_2 has been in recent years used as a structure-forming agent in petroleum and synthetic oils, i.e., as a thickener to produce lubricating greases [54-58]. It is dispersed under conditions excluding contact with air, in the medium of an organic liquid having a low boiling point, a low viscosity and a low surface tension.

Amines, phenolates, and carboxylates of alkali metals or zinc are added in order to reduce the corroding action of MoS_2. The specific surface of oleophilic MoS_2 amounts to 10-400 m^2/g. Taken in an amount of 10-40 %, it thickens well petroleum and synthetic oils, forming structurized systems, greases, offering high viscoelastic, antiwear and antiseizure properties. On dispersing and oleophilizating, MoS_2 does not lose its initial properties.

The MoS_2 consumption in production of lubricants is steadily rising [59-61]. Up to 80 thousand tons of MoS_2-containing greases are annually consumed all over the world for lubrication of friction assemblies of transportation facilities alone [61].

Molybdenum disulfide is commercially produced to TU 48-19-133-75 in four grades: DM-1, DM-2, DM-3, and DM-4 (Table 4.5). The stock for their production is the flotation molybdenum concentrate of grades KFM-1 and KFM-2 (GOST 219-69).

Graphite [52, 62, 63] is an allotropic form of carbon, whose layered crystal lattice consists of infinite planar parallel layers formed by regular hexagons of carbon atoms. Every carbon atom is surrounded with three adjacent ones at a distance of 1.42×10^{-5} m (1.42 A), to which it is bonded by very strong covalent bonds. The layers are spaced from one another at 3.65×10^{-5} m (3.65 A) [52]. The atomic layers in graphite are coupled by easily mobile electrons that are not attached to individual atoms, but move freely between the atomic layers. Such a bond is called metallic, since it determines the characteristic properties of metals. The thermal and electrical conductivities of graphite are of the same order as those of metals. The theoretical density of graphite is of 2270 kg/m^3; the Mohs hardness number along the C-axis is 5.5 and above; graphite is superior to molybdenum disulfide in thermal stability; its melting temperature is 3859±50 °C; sublimation temperature, 2700°C; and friction coefficient, less than 0.1 [62].

Table 4.6. Characteristics of graphite produced according to GOST 8296-73

Characteristics	*Graphite grade*				
	GS-1	GS-2	GS-3	GS-4	P
Ash content, % (not over)	0.5	1.0	2.0	5.0	7.0
Yield of volatiles, % (not over)	0.5	—	—	—	1.0
Content, %:					
moisture	0.5	0.5	0.5	1.0	1.0
sulfur	—	—	—	—	0.2
pH of aqueous extract	neutral	5.4-8.5	5.4-8.5	—	neutral
Residue, % (not over):					
on screen No. 020	—	—	—	0.05	absent
on screen No. 16	—	—	—	1.20	1.5
on screen No. 0063	—	1.0	1.0	—	—
Content of particles with equivalent sphere diameter more than $6 \cdot 10^{-6}$ m, % (not over)	40	—	—	—	—
Content of carbon, % (not over)	—	—	—	—	92

Graphite features a low friction coefficient, an exceedingly easy polishability, and plasticity. These properties show themselves more with higher degree of dispersity of graphite and lower orderliness of the arrangement of the crystals.

Graphite cleavage surfaces feature a considerable unsaturation of molecular attraction forces. Under the action of the forces, graphite adheres strongly to solid materials; this determines its antifriction properties: it fills all irregularities on rubbing surfaces and covers the latter with a continuous film. The friction on graphite surfaces being insignificant, this provides the efficiency of its lubricating effect. Antifriction properties of graphite stem also from the presence of an adsorption film on its surface [52], which in the air is always covered by an oxide layer. Substances adsorbed from the surrounding gaseous or liquid medium are present on the surface.

Graphites are produced to several GOSTs. Practically important as a grease component is graphite that meet the requirements of GOST 8295-75; their characteristics are presented in Table 4.6.

Graphite is inferior to molybdenum disulfide in the lubricating effect, but, owing to its ready availability and relative cheapness, is widely used in production of lubricants.

In grease production, graphite is primarily used as an antifriction addition [11,19]. Oleophilic graphite is employed as a thickener.

Mica [64-46] is natural and synthetic. Natural micas are aluminosilicate minerals with a layered structure, whose crystals have a shape close to hexagonal, with continuous layers of tetrahedrons. Micas are divided into several groups: (1) phlogopite $KMg_3[AlSi_3O_{10}] \cdot (OH,F)_2$ and muscovite $KAl_2[AlSi_3O_{10}] \cdot (OH)_2$, exhibiting high electrically insulating properties, which are produced in the form of a sheet mica and micaceous powder; (2) lepidolite $KLi_2Al[Si_4O_{10}] \cdot (OH, F)_2$ and cinvaldite $KLiFeAl[Si_3AlO_{10}] \cdot (OH, F)_2$, a starting stock for production of lithium and its salts; and (3) hydromuscovite $K_4Al_2[AlSi_3O_{10}] \cdot (OH)_2 \cdot nH_2O$ and vermiculite $(Mg^{2+}, Fe^{2+}, Fe^{3+})_3 \cdot [Si_4Al_4O_{10}] \cdot (OH)_2 \cdot nH_2O$, so-called hydromicas. On heating, their volume rises sharply (they bulge) because of liberation of water between silicon-oxygen packages.

Powdered phlogopite, muscovite and vermiculite (bulged) micas are more often used as fillers for sealing greases. A synthetic mica, produced by melting of MgO, Al_2O_3, SiO_2 together with metal fluorides (KF, MgF_2, AlF_3), can apparently also be used for this purpose. Characteristics of natural and synthetic micas are presented in **Table 4.7** [66]. Powdered mica produced to GOST 10698-80 and intended for rubber engineering components is more often used as filler for thread-sealing greases. Mica is also produced to GOST 855-74.

Boron nitride (BN) [67, 68]. There exist two boron nitride forms. An a-form of a graphite-type hexagonal structure develops under ordinary conditions; at 1360°C and a pressure of 6200 MPa it transforms into a b-form, known under the name of borazon, which exhibits the properties of diamonds. The a-form of boron nitride can, of course be, used as an addition or filler for greases. Boron nitride, grouped with covalent nitrides, is produced by action of nitrogen or ammonia on finely dispersed powdered boron or by reduction of boron oxide in presence of carbon: $B_2O_3 + 3C + N_2 = 2BN + 3CO$.

Table 4.7. Characteristics of natural and synthetic micas

Characteristics	Natural		Synthetic-fluoro-phlogopite
	muscovite	phogopite	
Density, kg/m^3	2600-2800	2300-2800	2600-2800
Water absorption, %	0.3-4.5	1.5-5.2	0.4-2.0
Thermal stability, °C	400-700	200-800	1000-1100
Cleavage force, MPa	0.14	0.17	0.17-0.20
Permittivity	7.2-7.5	5.5-6.3	6.7
Bulk resistivity at 20°C, ohm•m	10^{12}-10^{14}	10^{11}-10^{13}	10^{14}-10^{15}
Breakdown voltage at 20°C, kV	4	5	5.5

Boron nitride exhibits high resistance to oxidation, action of hot acids and other corrosive media, high (over 1000°C) thermal stability, and good lubricity.

Polymers [11, 19, 69]. Depending on the starting stock, production method and final product properties, products of polymerisation of various compounds polymers are and used in grease production either as the dispersion medium, or as the thickener, *i.e.*, the dispersed phase, or, sometimes, as viscous or adhesive additives. Some polymers can be employed as additions or – fillers, improving the lubricity of greases. The most efficient polymeric filler, which greatly upgrades the lubricity of greases, is powdered, polytetrafluoroethylene (PTFE), produced to GOST 10007-80 under the name of Fluoroplastic-4 in several grades and to GOST 6-05-402-80 as Fluoroplastic-40 in several grades as well. Such polymers are produced in the USA under names of Teflon, Halon; in Great Britain, Fluon; in Germany, Chostaflon; in Italy, Algoflon, etc. Polytrifluoroethylene (PTrFE), polytrifluorochloroethylene (PTFCE), etc. can also be used as fillers. Fluoroplastics and other halogen derivatives of polyethylene exhibit a high thermal stability, resistance to corrosive media and good low-temperature characteristics. They not only upgrade antiseizure and antiwear properties of greases, but also significantly improve their lubricity, preventing thereby an irregular (intermittent) sliding at low shear rates between rubbing surfaces. These polymers are also used as thickeners to produce special purpose greases. Characteristics of commercially produced halogen derivatives of polyethylene are presented in Table 4.8.

Apart from polytetrafluoroethylene and other halogen derivatives of polyethylene, polyethylene and polypropylene, which at their melting temperature dissolve well in oils and serve in this case as viscous additives can also be used as fillers. When added to greases in the form of powder at a temperature below the melting point of the polymer (60-90°C), they serve as fillers improving the antifriction properties of greases. Such fillers are employed for greases whose upper temperature limit of serviceability is not over 120°C.

Polymeric fillers are used in the form of powders with a particle size of 0.1-10 mm; they are added to greases in an amount of 1-5% at the stage of grease cooling before homogenization.

Metal oxides and metallic powders [8, 11, 19, 52], whose use was known over 100 years ago, are employed as fillers, especially in sealing greases. They are added not only to improve the anti-seizure and antiwear properties of greases, to prevent pitting but also to endow them with electrical conductivity. The use of these fillers, especially metallic powders, is efficient in case of sliding friction, but harmful in rolling friction. It should be also remembered that highly dispersed lead and copper powders are oxidation catalysts; lead is toxic; that other metals, contacting a free alkali in soap greases displace hydrogen, etc. It is very difficult to distribute any metal powder uniformly throughout the grease, which necessitates the use of special equipment to produce such greases.

Table 4.8. Characteristics of some halogen derivatives of polythylene

Characteristics	PTFE	PTFCE	PTrFE	PVC
Density, kg/m^3	2150-2250	2080-2160	1980-2000	1380-1400
Temperature, °C:				
melting	327	210-225	195-210	190-198
decomposition	509	411	413	389
Brinell hardness, MPa	30-40	100-130	24-25	100-130
Abrasion resistance, m^2 per 1000 cycles	—	4.8-7.5	—	23
Friction coefficient	0.05	0.3	0.14-0.17	0.15-0.30
Operating temperature, °C:				
maximum	260	150	150	110-150
minimum	-269	— 196	— 60	— 70
Combustibility	Noncombustible	Noncombustible	Noncombustible	Self-extinguishing
Fungi and tropical stability	Good	Good	Good	Good

The metal oxides and metallic powders, most widely used as grease fillers, include zinc oxide, titanium oxide, powders of aluminium, copper, lead, tin, bronze and brass; they are added to a finished grease by mixing in amounts of 1-5% and sometimes, of 15-30%.

REFERENCES

1. Additives to Oils. Proc. of 2nd All-USSR Sci. and Techn. Meet. Ed. by Krein S.E., Sanin P.I., Kuliev A.M. and Eminov E.A., Khi-miya, Moscow, 1966; Ibid., 1968.
2. Kuliev A.M., Chemistry and Tecnology of Oil and Fuel Additives, Khimiya, Moscow, 1972.
3. Vipper A.B., Vilenkin A.V. and Gaisner D.A., Foreign Oils and Additives, Khimiya, Moscow, 1981.
4. Vinogradova I.E., Antiwear Oil Additives, Khimiya, Moscow, 1972.
5. Zaslavsky Yu. S. and Zaslavsky R.N., Mechanism of Action of Antiwear oil Additives, Khimiya. Moscow, 1978.
6. Shekhter Yu. N. and Krein S.E., Surfactantsfrom Petroleum Stock, Khi-miya, Moscow, 1971.
7. Shekhter Yu. K., Krein S.E. and Teterina A.N., Oil-Soluble Surfactants, Khimiya, Moscow, 1978.

ADDITIVES AND FILLERS

8. Velikovsky D.S., Oil and Grease Additives Based on Petroleum Hydrocarbon Oxidation Products, GOSINTI, Moscow, 1960.
9. Boner K.D., Production and Application of Greases. Transl. from English, Ed. by Sinitsyn V.V., Gostoptekhizdat, Moscow, 1958.
10. Velikovsky D.S., Poddubny V.N., Vainshtok V.V. and Gotovkin B.D., Consistent Greases, Khimiya, Moscow, 1966.
11. Sinitsyn V.V., Plastic Lubricating Greases in the USSR (Reference Book), Khimiya, Moscow, 1984.
12. Vainshtok V.V., Fuks I.G., Shekhter Yu.N. and Ishchuk Yu.L., Composition and Properties of Plastic Lubricating Greases (Review), TsNIITEneftekhim, Moscow, 1970.
13. Papok K.K. and Ragozin N.A., Dictionary of Fuels, Oils, Greases, Additives, and Special Liquids, Khimiya, Moscow, 1975.
14. Ishchuk Yu.L. and Fuks I.G., Modern State and Prospects of Research in Field of Plastic Greases, in: Plastic Greases, Proc. of 2nd All-USSR Sci. and Techn. Conf. (Berdyansk, 1975), Naukova Dumka, Kiev, 1975, pp. 7-11.
15. Fuks I.G., Research and Development of Plastic Lubricating Greases with Additives-Fillers, Dr. Tech. Sci. Thesis, Moscow, 1979.
16. Lendyel I.V., Nedbailyuk P.E., Shkolnikov V.M. et al., Effect of Formulation-and-Production Process Factors on Low-Temperature Properties of Greases, in: Plastic Greases, Proc. of 2nd All-USSR Sci. and Techn. Gonf. (Berdyansk, 1975), Naukova Dumka, Kiev, 1975, pp. 120-121.
17. Fuks I.G., Uvarova E.M., Vdovichenko P.N. and Kitashov Yu.N., Improvement of Lubricity of Plastic Greases with Compositions of Additives and Fillers, Neftepererabotka i Neftekhimiya, Kiev, 1981, Issue 20, pp. 23-28.
18. Kravchenko A.R., Development of Method and Investigation into Protective Properties of Greases under Dynamic Conditions, Cand. Tech. Sci. Thesis, Moscow, 1982.
19. Fuks I.G., Additives to Plastic Lubricating Greases, Khimiya, Moscow, 1982.
20. Plastic Greases. Abstr. of Reports at 1st, 2nd, and 3rd All-Union Sci. and Techn. Conf., Naukova Dumka, Kiev, 1971, pp. 134-173; Ibid., 1975, pp. 56-91; Ibid., 1979, pp. 67-88.
21. MacDonald R.T. and Dreher I.L., Additives for Greases, Oil and Gas J., 1953, Vol. 3, No. 23, pp. 34-33.
22. Silver H.B. and Stanley I.R., The Effect of Thickeners on the Efficiency of Load-Carrying Additives in Greases, Tribology Int., 1974, No. 6, pp. 113-118.
23. Mankovskaya N.K., Modern Notions of Chemism of Processes of Oxidation of Various Hydricarbon Classes, Its Inhibition, and Oxidation Stability of Plastic Lubricating Greases, Naukova Dumka, Kiev, 1976.
24. Butovets V.V., Development of Method and Investigation into Kinetics of Oxidation of Plastic Lubricating Greases, Cand. Chem. Sci. Thesis, Kiev, 1976.
25. Trites R.T., Oxidation Inhibitor for Extreme Temperature Greases, NLGI Spokesman, 1968, Vol. 32, No. 3, pp. 168-174.
26. Pat. 2, 921, 899 (USA). Oxidation-Resistant Lubricating Greases Containing Inorganic Alkali Metal Compositions of High Alkalinity. Z.W.Sproule, J.H.Norton, W.C.Pattenden. Publ. Jan. 19, 1960.
27. Pat. 2, 951, 808 (USA). Lubricant Compositions Containing Metal Salts of Aromatic Hydroxy Carboxylic Acids as Antioxidants. J.H. Norton, W.C.Pattenden. Plibl. Sept. 6, 1960.

28. Pat. 3, 000, 819 (USA). Finally Derived Metal Salts as Antioxidants for Oils and Greases. J.H. Norton, Z.W. Spoule. Publ. Sept. 19, 1961.
29. MacDonald R.T. and Dreher J.L., Additives for Lubricating Greases, NLG-I Spokesman, 1953, Vol. 1, No. 1, pp. 16-26.
30. Zakin I., Lin H. and Tu E., Exploratory Studies of the Sorption and Extraction of Additives in Lubricating Greases, NLGI Spokesman, 1967, Vol. 31, No. 2, pp. 43-48.
31. Murphy G.JP., Factors that Influence Grease Oxidation and Oxidative Wear, Ibid., 1964, Vol. 28, No. 1, pp. 15-21.
32. Patzau S., Oxydationsstabilitat der Schmierfette, Ereiberger Porschungsh., 1965, Bd. A367, S. 235-241.
33. Novoded R.D., Investigation into Thermooxidation Stability of Complex Calcium Greases by Thermal Analysis and IR Spectroscopy Methods, Cand. Tech. Sci. Thesis, Moscow, 1980.
34. Bish I.M., Investigation of Extreme Pressure Additives for Greases, NLGI Spokesman, 1970, Vol. 33, No. 12, pp. 426-434.
35. Kuzmichev S.P., Popova I.E. and Ishchuk Yu. L., Hypoid Additives to Lithium Oxystearate-Based Greases. Neftyanaya i Gazovaya Promyshlennost, 1974, No. 2, pp. 36-38.
36. Mankovskaya N.K., Ishchuk Yu. L., Popova I.B. et al., Sulfur-and Lead-Containing Additives for Lithium Stearate-Based Greases, Neftepererabotka i Neftekhimiya, 1975, No. 9, pp. 14-16.
37. Calhoum S..F., Antiwear and Extreme Pressure Additives for Greases, ASLE Trans., 1960, Vol. 3, No. 2, pp. 208-214.
38. Vainshtok V.V., Smirnova N.S. and Vavanov V.V., On Effect of Antiwear and Antiseizure Additives on Properties of Lithium Greases, Neftepererabotka i Neftekhimiya, 1975, No. 4, pp. 20-22.
39. Zhdanov I.P., Podolsky Yu.Ya. and Zurkan I.G., On Efficiency of Action of Antiwear Additives in Plastic Lubricating Greases for Axle Boxes of Railway Cars, Khimiya i Tekhnologiya Topliv i Masel, 1977, No. 4, pp. 50-54.
40. Shekhter Yu.N., Shkolnikov V.M., Bogdanova T.I. and Miovanov V.D., Working-and-Preservation Lubricants, Khimiya, Moscow,1979.
41. Kosarskaya Yu.P., Protective Properties of Plastic Lubricating Greases and Ways for Their Improvement, Cand. Tech. Sci. Thesis, Moscow, 1974.
42. Altsybeeva A.I. and Levin S.Z., Metal Corrosion Inhibitors, Khimiya, Moscow, 1968.
43. Friberg S., Rudhag L. and Atterby P., Analytic Methods for the Estimation of the Influence of Corrosion Inhibitors on the Properties of Lithium Soap Grease, Tribology, 1970, Vol. 3, No. 1, pp. 39-42; No. 4, pp. 236-237.
44. Young R.L., Rust Preventive Abilities of Grease and Their Improvements, Lubr. Eng., 1963, Vol. 19, No. 7, pp. 293-296.
45. Rhodes R.K., Petroleum Sulfonates in Grease Manufacture, Inst. Spokesman, 1951.
46. Sinitsyn V.V., Mikheev E.S., Tseptsova V.M. and Grishin N.N., Evaluation of Preservation Properties of Plastic Lubricating Greases with the Aid of Multielement Sensor, Khimiya i Tekhnologiya Topliv i Masel, 1981, No. 2, pp. 45-46.
47. Sinitsyn V.V., Grishin N.N., Gvozdilina V.M. and Utekhina S.N., Improvement of Preservation Properties of Multipurpose Lithium Greases, Ibid., No. 11, pp. 37-38.
48. Samgina V.V., High Polymers in Plastic Lubricating Greases, Cand. Tech. Sci. Thesis, Moscow, 1974.

49. Nikolaev A.F., Synthetic Polymers and Plastics Based on Them, Khimiya, Moscow-Leningrad, 1964.
50. Shchegolev G.G., Microstructure and Properties of Lithium Greases, Cand. Chem. Sci. Thesis, Moscow, 1968.
51. Ishchuk Yu. L., Investigation into Influence of Dispersed Phase on Structure, Properties, and Technology of Plastic Lubricating Greases, Dr. Tech. Sci. Thesis, Moscow, 1978.
52. Braithwaite E.R., Solid Lubricants and Antifriction Coatings, Khimiya, Moscow, 1967.
53. Troyanovskaya G.I., Solid Lubricants and Coatings for Operation in Vacuum, in: Friction, Wear, and Lubricants (Reference Book), Mashinistroenie, Moscow, 1978, Vol. 1, pp. 283-295.
54. Pat. 1, 162, 222 (Great Britain). Improved Dispersions and Greases. A.J.Groszek. Publ. Sept. 20, 1969.
55. Pat. 1, 296, 904 (Great Britain). Metal Sulphide/Graphite Composition. A.J.Groszek, Ch. J. Grach. Publ. Nov. 22, 1972.
56. Pat. 1, 169, 784 (Great Britain). Improved Greases and Dispersions. J.Woulard. Publ. Oct. 29, 1969.
57. Pat. 30, 811 (Ireland). Improved Dispersions and Greases. Publ. Jan. 26, 1972.
58. Pat. 3, 518, 191 (USA). Modified Oleophilic Graphite and Heavy Metal Sulphides. A.J. Groszek, G.H.Bell, C.D.Staines. Publ. Feb. 21, 1968.
59. Barry H.F., Factors Relating to the Performance of MoS_2 as a Lubricant, Lubr. Eng., 1977, Vol. 33, Ho. 9. pp. 476-480.
60. Risdon T.J., Some Energy Implications for the Use of MoS_2 in Greases, NLGI Spokesman, 1977, Vol. 41, No. 5, pp. 150-159.
61. Braithwaite E.R. and Greene A.B., A Critical Analysis of the Performance of Molybdenum Compounds in Motor Vehicles, Wear, 1978, Vol. 46, No. 2, pp. 405-432.
62. Industrial Requirements for Mineral Stock Quality. Reference Book for Geologists. Issue 3. Graphite, Gosgeoltekhizdat, Moscow, 1960.
63. Shulepov S.V., Carbon Atom and Synthetic Graphite, Yuzhno-Ural. Kh. Izdat., Chelyabinsk, 1965.
64. Lashev E.K., Mica. Evaluation of Deposits in Surveying and Prospecting (Guide for Geologists), Gosgeolizdat, Moscow, 1948, Issue 1.
65. Leizerzon M.S., Synthetic Mica, Gosenergoizdat, Moscow-Leningrad, 1962.
66. Tresvyatsky S.G., Micas, in: Brief Chem. Encyclop., 1965. Vol. 4, pp. 914-916.
67. Samson G.V., Kitrides, Ibid., 1964, Vol. 3, pp. 490-494.
68. Organic Borates, in: Chemical Encyclop. Dictionary, 1983, p. 79.
69. Panshin Yu.A., Malkevich S.G. and Dunaevskaya Ts.S., Fluoroplastics, Khimiya, Leningrad, 1978.

Characteristic of Lubricating Grease Production Technology. Its Main Stages and Their Significance in the formation of Structure and the Properties of Greases

Specific features of grease production processes and the equipment used for them are primarily determined by the nature of the thickener [1-9]. The simplest is the production technology for hydrocarbon greases, which consists of a preliminary preparation of the starting components (solid hydrocarbons and oils), their melting together, dewatering and subsequent cooling of the melt. In production of greases thickened with highly dispersed hydrophobized silica, organoclays, oleophilic graphite, carbon black, pigments and other inorganic and organic thickeners (except with urea derivatives produced in the course of the grease preparation, i.e., in situ) the main production process stage consists of a mechanical dispersion of the thickener in an oil and a subsequent homogenisation. The production technology for soap greases, thickened with soaps, as well as greases thickened with urea derivatives produced in situ is a complex and multistage process and calls for special equipment.

The process of production of soap-thickened greases consists of the following main stages: preparation and proportioning of the stock; production of the thickener; its thermomechanical dispersion; cooling of the melt; homogenisation; filtration; deaeration and packing [1-7,10-16].

Stock proportioning. The proportion of the components of a grease determines its properties. Thus, increasing the thickener concentration significantly increases the shear strength, viscosity of the grease, reduces the penetration and the oil separation under pressure [1, 2, 7, 17]. Charging an excess or a lack of alkali as against the stoichiometrically needed amount gives rise to free acids or alkalis in the grease and changes considerably its structure and properties [1, 2, 7, 15-24]. For example, the variation of rheological characteristics of calcium (hydrated and complex) and lithium hydroxystearate greases with the concentration of free acids and alkalis is of extreme importance (Fig. 5.1) [21-24]. The highest values of shear strength and viscosity and the lowest oil separation under pressure are characteristic of neutral or weakly alkaline greases. Increase in the grease alkalinity or acidity greatly lowers the thickening effect of a soap as well as impairs other performance characteristics of greases. Thus, an excess of a free acid (over 1 %) in lithium greases increases the thermal hardening and impairs their thermal stability [21-23]. This is also valid for other grease types, being particularly pronounced for cCa-greases [17-24]. Neutral and, to a greater extent, acidic cCa-greases distinctly exhibit an anomalous temperature dependence of the shear strength: at 50°C it is higher than at 20° and 80°C; this adversely affects the grease behaviour in a lubrication point. Neutral and acidic cCa-greases are also more prone to surface hardening and volume compaction in storage than alkaline ones. The above considerations necessitate the maximum accuracy of the stock propoptioning in grease production. A particularly high accuracy of proportioning should be provided in continuous processes, where, in contrast to batch production methods, it is difficult to correct the proportion of ingredients in the course of the process.

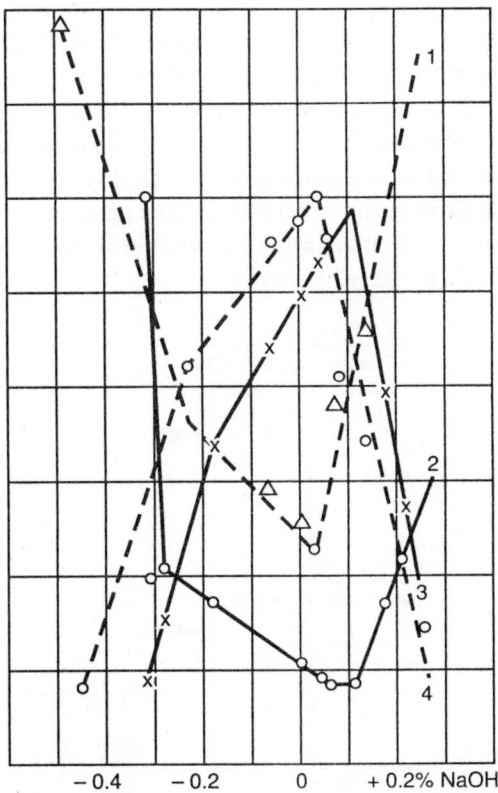

Fig. 5.1. Effect of concentration of free acids and alkalies on colloidal stability (1,2) and shear strength (3,4) of greases based on calcium stearate and oleate (2,3) and on lithium oxystearate (1,4).

Liquid starting components are commonly metered at 20-85°C. The charging temperature exerts no substantial effect on grease properties. Volume batching devices are more often used in batch and semicontinuous processes; an inadequate accuracy of their operation is compensated for by correcting the ratio of ingredients and mixing them in the reactor. Apart from the volume batching, weight batching also is used, for which purpose the batching tanks are installed on scales or on devices fitted with strain gauges. A reactor-stirrer where components are mixed and in production of soap greases, also wherein acids (fats) are neutralized (saponified), is often installed on a device with a strain gauge; the advantag being that such a method provides a high accuracy of the stock proportioning.

A wide use has been found of metering pumps with adders, which automatically maintain the required proportion of ingredients [25]. Metering pumps are mandatory element of the production line in continuous processes. They are capable of metering not only oils and other liquids, but also suspensions of free-flowing materials in a dispersion medium, such as of hydrated lime in oil. Some metering pumps are equipped with electric actuators for a remote and an automatic control of the metering feed. Serially produced metering units of the DA series, including several metering pumps of the required deliveries, are most suitable for use in continuous processes.

The order (sequence) of charging of the components of a soap thickener is essential in some cases. Thus, the sequence of charging of the ingredients forming an aluminium complex thickener significantly affects the properties of cAl-greases [26]. Figure 5.2 shows schematically three versions of cAl-grease production technology, differing in the sequence of charging of the stock components (AIP-aluminium isopropylate; HSt-stearic acid; HBz-benzoic acid); Table 5.1 presents the properties of the resulting greases. As can be seen, production of cAl-greases by version II yields the best results. It should be noted that the sequence of charging of components affects the properties in production of other complex greases: lithium, calcium and barium. This is to be taken into account and the component charging order specified by the production technology procedure should be strictly observed.

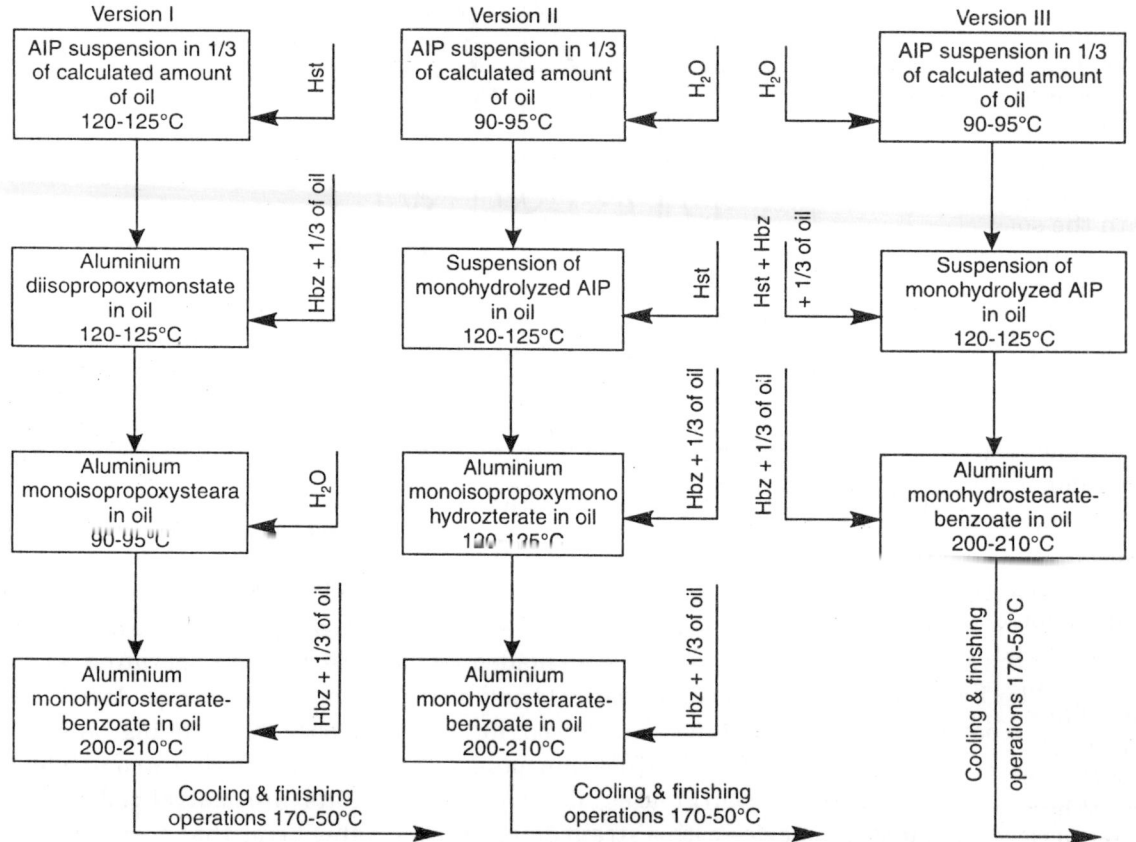

Fig. 5.2. Versions of different sequences of charging of components in manufacturing cAL-greases (total consumption of HSt + HBz = 10%).

Preparation and thermomechanical dispersion of thickener. This is an important stage of the production process [1-9, 15-17]. Saponification of fats or neutralization of fatty acids is the initial stage of the basic process of production of soap greases. The completeness of saponification of fats is of great importance for producing greases having predetermined properties with minimum consumption of fatty materials. Saponification of fats or neutralization of acids is effected by an aqueous solution of alkalis or by their oil suspensions and pastes in presence of some amount (usually, 1/3-2/3 of the calculated amount) of oil, the dispersion medium of greases. This accelerates the reaction of saponification and facilitates the dispersion of soap in oil at subsequent process stages. Soaps of monovalent metals as well as of calcium and barium are obtained by direct saponification of fats with alkalis. Production of greases based on soaps of polyvalent heavy metals-zinc, lead, etc.-involves a double exchange reaction: first the reaction of saponification of fats with the hydroxide of a monovalent metal (mainly, sodium) and then the reaction between the sodium soap and a water soluble salt of the corresponding polyvalent metal in an aqueous medium are conducted. The soap of a polyvalent metal is practically water insoluble and precipitates at the exchange reaction. It is then washed off from water soluble salts and used for the grease production in a moist state or, after drying, as a powder; lead soaps are used after melting and a corresponding dispersion of the soap. Aluminium soaps (especially complex ones), stable and suitable for production of high-quality greases, are prepared, with the use of aluminium alcoholates, most often of aluminium isopropylate (AIP).

After the saponification process, moisture or the excess of alcohol is fully (for hydrated calcium, calcium-sodium, and sodium greases, up to a certain limit) removed from the soap-oil suspension. In production of greases with finished air-dry soaps, the soap-oil suspension is obtained by directly mixing components in the specified proportions, after which it is heated till a homogeneous isotropic melt is obtained. Producing finished air-dry soaps beforehand and then using them in production of greases is, however, economically disadvantageous as in this case the total energy consumption is 20-25% higher than that required for producing greases with the soap preparation in situ. Moreover, for cAl-greases the use of finished air-dry soaps (moisture content, upto 1%; free acidity, 4-6%; and ash, 11-12%) at a total consumption of acids of 10% for the finished grease significantly impairs its properties (penetration, 260-280; shear strength at 20°C, 450-520 Pa; colloidal stability, 7-9%) as compared with the grease produced by version II (see Fig. 5.2 and Table 5.1).

Grease production methods are also known where the soap-oil emulsion is not brought to the isotropic melt formation temperature, but only swelling of the soap in the oil at a relatively low temperature is provided [1, 2, 5, 27]. Such a method has been called "cold mixing" or low-temperature production process.

The soap preparation stage is a very important step in the grease production process. Special attention should be paid to the soap preparation with the use of animal or vegetable fats. In this case the saponification of fats is the most energy-intensive procedure, involving destruction of triglycerides of saturated (mainly stearic) and unsaturated (mainly oleic) acids. An incomplete saponification of fats reduces the soap concentration and results in presence of free fats in the grease, which impairs the grease properties and entails additional consumption of fats. The simplest method for cutting down the energy consumption at this stage is the use of autoclaves. The autoclaving should be carried out not only with a preliminary steam fed into the autoclave jacket, but also with steam fed directly into the autoclave [28]. In production of hydrated calcium greases this method shortens the soap concentrate preparation cycle by more than twice as compared to an ordinary method where the pressure rise in the autoclave is attained solely by heating its contents to 130-135°C. Apart from shortening the soap preparation cycle, the steam fed directly into the autoclave yields fats with better degree of saponification, which provides for savings of fats. For example ordinary autoclaving the hydrolysis of fats after 3 h of saponification is 90-94%, while with additional steam fed directly into the autoclave the saponification degree is as high as 95-97% after 1.5 h under the same conditions [28]. This provides for further shortening of the cycle of grease production in open vessels, where the steps after saponification of fats and correction of the free alkali in the grease are effected. Moreover, after saponification of fats, water should be fed into an open vessel not intermittently, i.e., by portions (Fig. 5.3, curve 1), but continuously at a rate providing for constancy of the optimum saponification temperature (Fig. 5.3, curve 2); this excludes inactive zones and shortens the fat after saponification process time by about 1.5 times one third.

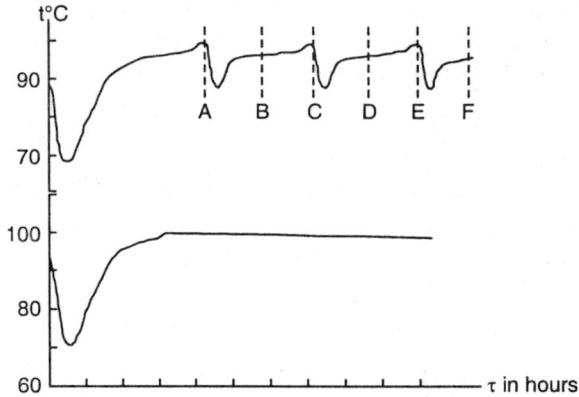

Fig. 5.3. Temperature conditions of saponification of fats at different methods of water feed into reaction zone in manufacture of hydrated calcium greases [28] : τ, h
1-intermittent (by portions) water feed (AB, CD, EF, -inactive zones);
2-continuous water feed; A, B, C, D, E, F.

TABLE 5.1. Characteristics of aluminium complex greases manufactured at different component charging sequences

Characteristics	Process versions in Fig.5.2		
	I	II	III
Penetration at 25°C, 10-1 mm	220	190	310
Shear strength, Pa: at 20°C at 80°C	 1020 470	 1480 610	 240 90
Viscosity at 0°C and 10 s-1, Pa·s	320	440	100
Colloidal stability, %	2.9	2.5	12.5

Note. The drop points of all greases exceeded 250°C.

Grease properties are markedly affected by the rate of the thermomechanical dispersion of soap with its heating to the isotropic melt temperature as well as by the melt holding time at this temperature [1,2,6,8,15,17,20,29-50]. In production of sodium and lithium greases the temperature rise rate at the thermo-mechanical treatment stage should be as high as possible; this, firstly, shortens the grease preparation cycle time and, secondly, reduces the evaporation loss as well as the oxidation degree of the dispersion medium. In production of, e.g., cCa-greases, however, the optimum thermal treatment rate should be maintained since the formation of an adsorption soap complex occurs at this stage: calcium acetate "builds in" itself into the crystal lattice of the soap from the moment of water removal from the system and upto the thermal treatment temperature.

The optimum heating rate for Uniol-type cCa-greases, i.e., those prepared with the use of C10-C20 SFA, was ascertained [8,17, 29] to be of 1-1.5 deg/min. In addition, when producing complex calcium greases, it is expedient to effect at the heating stage an isothermal treatment of the soap-oil system at 160-170°C for 20-25 min since an intense interaction between calcium acetate and stearate occurs at this temperature [30]. Such a production technique upgrades the thickening effect of the soap and stabilizes the properties of cCa-greases with time [30].

An important factor of the stage of soap preparation and thermomechanical treatment of the soap-oil system is the final temperature of the thermal treatment and holding time at the temperature [6-8, 28-37]. This stage is of particular importance in production of complex-soap greases. The influence of the maximum heating temperature and of its duration on properties of complex calcium greases prepared with C10-C20 SFA and acetic acid is evidenced by the data of Table 5.2.

TABLE 5.2. **Dependence of properties of complex calcium greases on thermal treatment temperature and melt holding time**

Final heating temperature, °C	Holding time min	Shear strength at 50°C, Pa 10 Pa·s	Viscosity at 50°C and 1000 s-1	Oil separation under pressure %
180	10	360	22	10.6
200	10	440	32	4.9
200	60	640	49	3.0
225	10	700	60	4.2

Complex calcium greases prepared at temperatures up to 200°C exhibit low yield strengths and are unstable in storage. Even increasing the thermal treatment time to 1 h fails to yield the desired results. The optimum temperature of the thermal treatment of Uniol-type cCa-greases at a relatively short holding time (10-15 min) is 220-225°C [8,17,31-34]. These conditions provide for a high thickening effect of a cCa-soap in petroleum oils, good thermal, mechanical and thermomechanical stabilities of these greases as well as an adequate stability of their properties in storage. A higher thermal treatment temperature is also used, particularly when polysiloxanes or other synthetic heat-resistant liquids serve as the dispersion medium of complex calcium greases. Thus, grease TsIATIM-221 is prepared at 250°C [6].

The maximum temperature of preparation of hydrated calcium greases (cup greases) is determined by the soap concentrate treatment temperature ensuring the specified water content in the finished grease, which depends on the composition of the saponifiable stock (natural fats or synthetic fatty acids, etc.) and is within 105-110°C. For fatty cup greases, i.e., those produced with natural fats, it is somewhat higher than for synthetic ones.

The authors of [8, 35-38] propose the thermal treatment of anhydrous calcium greases to be conducted at 120-165°C. The optimum thermal treatment temperature is, however, determined by the dispersion medium composition and saponifiable components of the grease. When producing anhydrous calcium greases from petroleum oils and 12-hydroxystearic acid, the soap-oil suspension should be heated at a rate of about 1 deg/min up to 120-125°C (mainly 120°C) which corresponds to the temperature of the phase transition of calcium 12-hydroxystearate to a liquid-crystalline state and to the maximum rate of gelation in the system [8, 37, 38]. The grease is to be held for about 1 h at this temperature.

The optimum temperature of the thermal treatment of lithium greases would be that temperature at which an isotropic melt of the soap in the oil is attained (about 205-210°C) [1, 2, 15, 20, 39, 40]. To avoid significant evaporation losses of the dispersion medium as well as to reduce its oxidation, the melt holding time at this temperature should not exceed 10 min. The same thermal treatment temperature and time are also recommended for other types of high-temperature soap greases, such as aluminium complex greases [41, 42]. The time of holding of cAl-greases at this temperature should not exceed 30 min. The so-called low-temperature method of preparation of lithium greases [5, 27] is employed sometimes, chiefly in production of lithium hydroxystearate greases. It involves a thorough dispersion of the soap in the dispersion medium at every production process stage with the use of efficient mixers and homogenizer valves. In the lowtemperature method of production of lithium hydroxystearate greases the dispersion thermal treatment temperature is not over 180-185°C.

Apart from ordinary lithium greases, lithium complex (cLi-) greases are also produced [43-47]. Greases prepared with three-component lithium soaps, whose composition includes lithium hydroxystearate, lithium salts of boric and terephthalic acids, are of the greatest interest. In production of such greases from petroleum oils, such as I-50A (LKS-type greases), first 12-hydroxystearic acid in the oil medium is neutralized with an excess amount of LiOH and then boric acid is charged at 100-105°C and terephthalic acid at 120-130°C. The suspension is heated to 230-240°C (preferably, 230°C) and held for about 30 min [44-46]. This, at a total complex lithium soap consumption of 10%, yields a grease with a penetration of $(205-270) \times 10^{-4}$ m, a shear strength at 20°C of 1000-1300 Pa, a drop point over 230°C and a colloidal stability of 6-8%. Lowering the thermal treatment temperature significantly impairs the grease properties: the shear strength and oil separation under pressure each decrease by a factor of 2 and the drop point decreases by 20-25°C. Increasing the temperature of thermal treatment of cLi-greases above 240°C as well worsens their physico-chemical and rheological characteristics.

The process of thermal treatment of complex barium (cBa-) greases consists of several temperature stages [48-50], whose choice depends on the nature of the dispersion medium and the composition of the main (high molecular) saponifiable component. Thus, when using C10-C16 SFA and other fractions of the acids, it is proposed to carry out the thermal treatment in two stages [50]. The reaction mixture is first heated to 125-135°C and then, after adding a certain part of the dispersion medium (petroleum oil), held at this temperature for about 1 h. Next, the heating is continued to 145-150°C and the mixture is once again held, now at this temperature, for 1.5-2 h. It is also recommended to carry out the thermal treatment of complex barium (cBa-) greases at higher temperatures and longer holding time [1,49]. Thus, the authors of [49] propose the thermal treatment of a cBa-grease to be carried out for 3-4 h at 205-215°C.

In the domestic practice of production of barium complex greases, such as ShRB-4, the soap-oil suspension is treated at 140-145°C for about 0.5 h. The required quality of sodium and sodium-calcium greases is provided by the thermal treatment of the soap-oil suspension at 140-160°C [1,2]. The selection of the optimum temperature in this case as well depends on the composition of the dispersion medium and saponifiable components.

Lithium-calcium greases are preferably thermally treated at 205-210°C for not more than 0.5 h, the suspension of soap in oil being heated, to this temperature at a rate of 1.5-2 deg/min [18].

The final temperature of thermal treatment is essential in production of some polyurea greases. Thus, for greases whose production involves the preparation of a polyurea thickener in situ in a petroleum oil by the interaction of diphenylmethanediisocyanate with a mixture of aliphatic primary amines (alkali number, 176; average molecular weight, 313) at final thermal treatment temperatures of 150, 160, 170, and 200°C the penetrations were of 330, 250, 210, and 205×10^{-4} m and the shear strengths at 20°C were of 160, 420, 870, and 1030 Pa respectively; the thickener consumption was 20%. A reduction of the thickener consumption to 15% and the thermal treatment at 200°C yielded a grease with a penetration of 220×10^{-4} m and a shear strength at 20°C of 550 Pa. The grease drop point was 250°C in all cases.

Production of greases with a finished polyurea, such as the product of interaction of diphenylmethanediisocyanate and aniline, requires, after a thorough homogenisation (shear rate, 5×10 s-1), a prolonged (2-3 h) thermal treatment at 180-200°C. Production of greases with the use of a polyurea acetate complex as well requires heating upto 190-200°C.

Cooling of the soap melt in oil is one of main stages of the soap grease production process [1-9, 15, 17, 19, 20, 29, 30, 35-40, 51-64]. As the soap melt in oil cools down, the grease skeleton is forming, which determines the main rheological and other properties of the grease. This process is a multistage one and proceeds in a very complicated manner. First the crystallization centers and micelles appear and then over-micellar crystal nuclei emerge and grow, and fibres develop, from which the three-dimensional grease skeleton is formed. Many of the stages proceed concurrently. It should be noted that the grease skeleton formation processes have been inadequately studied, so far. The crystallization of soaps in the dispersion medium and subsequent structure formation stages are affected by numerous factors: the composition of the dispersion medium; the nature and concentration of the soap; presence of free acids and alkalis; presence of other surfactants, such as additives, and their concentration; the rate of stirring of the crystallizing system; and, naturally, its cooling conditions. With unchanged compositions of the dispersion medium and dispersed phase, their accurate proportioning, and constancy of other process parameters, the conditions of cooling of the soap melt in oil are the governing factors in the formation of the grease skeleton. Varying the cooling conditions (rapid or slow cooling, isothermal crystallization), it is possible to act on the grease formation process so as to control the size and shape of dispersed phase particles (fibres) and the formation of the grease skeleton, i.e., to change the grease properties in the desired manner.

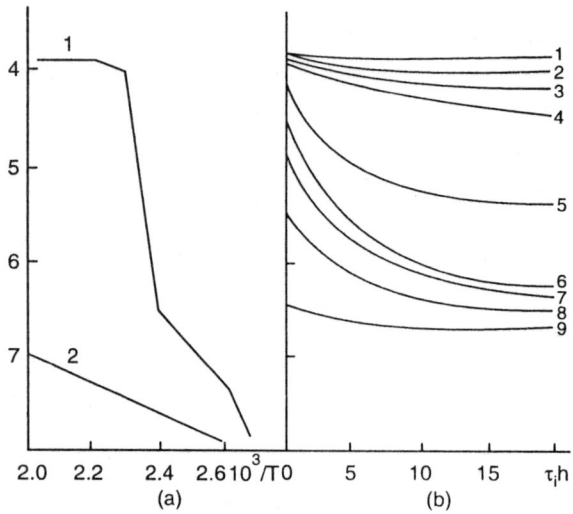

Fig. 5.4. Electrical conductivity of soap dispersion in oil as a function of temperature (a) and of time (b) within temperature range from 210 (1) to 140°C (9).

The influence of the stage of cooling of the soap melt in oil on its crystallization and for the grease formation process can be illustrated by the character of variation of the electrical conductivity of a 10-% dispersion of lithium 12-hydroxystearate in vaseline oil in the melt cooling and by isotherms of crystallization of the soap in the oil (Fig. 5.4) [51,52,55]. The temperature dependence of the electrical conductivity of the dispersion of the soap in the oil (Fig. 5.4,a, curve 1), in contrast to that of the electrical conductivity of the oil itself (curve 2), is of a complex nature. The electrical conductivity of the oil is very low (1×10^{-7} ohm-1); it is determined by presence of small amounts of current-conducting impurities and linearly decreases with temperature. The conductivity of the soap dispersion, in contrast, has distinct abrupt changes; several of its characteristic linear segments can be distinguished. The abrupt changes of the lg I(1/T) curve for the lithium hydroxystearate (LioSt) - vaseline oil system stem from the specifics of its formed structures on cooling. The steep decline of the electrical conductivity of the system with lowering temperature and with time at a definite temperature (Fig. 5.4, b) represents a total decrease of the concentration of the crystallizing substance (LioSt) in the solution regardless of whether it is spent for the formation of new crystallization centres or for the growth of already existing crystals, elements of the grease skeleton. The slope of the curves shows the rate of variation of the system's electrical conductivity with temperature (Fig. 5.4, a) and with time at a given temperature (Fig. 5.4, b) and hence also the crystallization rate, or the rate of passage of soap molecules from the solution to the surface of the solid phase. As can be seen, high crystallization rates are within a temperature range of 180 to 150°C. A more exact determination of the temperature at which the maximum rate of passage of soap molecules to the surface of crystals having formed will occur, i.e., the most rapid growth of the crystals will be provided, calls for evaluating the dependence of the electrical conductivity variation rate on the crystallization temperature. The curve presenting the dependence has a

Table 5.3. **Influence of cooling method on properties of lithium greases**
(I – equal-mass mixture of oils AU and I-50A, thickened with 2% polyisobutylene P-20 + 12% LioSt; II – oil ASV-5 + 10% LioSt)

Method	Cooling	Penetration at 25°C after 60-fold stirring, mm•10^{-1}		Shear strength, Pa		Viscosity at 0°C and 10 s^{-1}, Pa•s		Colloidal stability, %	
		I	II	I	II	I	II	I	II
1.	*Rapid.* On water-cooled mill rolls from 210 to 60-50°C at rate of 35-40 deg/min	280	—	460	—	210	—	—	—
2.	*Rapid.* On pan in thin layer from 210 to 60-50°C at rate of about 30 deg/min	270	—	500	840	210	150	14.8	15.5
3.	*Slow.* In kettle with cold oil feed into cooker jacket: from 210 to 120°C at rate of 1.5-2 deg/min and from 120 to 50-60°C at rate of 1-1.5 deg/min	200	—	1250	—	400	—	7.5	—
4.	*Combined.* In kettle from 210 to 165°C at rate of 1.5-2 deg/min by addition of dispersion medium to grease, followed by rapid cooling at rate of 20-25 deg/min in scraper-type cooler	—	—	—	980	—	200	—	13.0
5.	*Combined (stepwise).* In kettle from 210 to 185-187°C by addition of dispersion medium to grease (1.5-2 deg/min); holding for 30-40 min (isothermal crystallization); cooling to 165°C by coolant feed into cooker jacket (2-3 deg/min); and cooling to 60-50°C in scraper-type cooler at rate of 20-25 deg/min	—	—	—	1220	—	250	—	10.2

pronounced maximum corresponding to the temperature of the maximum crystallization rate, i.e., to the most rapid growth of crystals that are elements of the grease skeleton. Note that for the system under consideration (vaseline oil - LioSt) this temperature is close to 170°C [51-55]. For oil S-220 it is 160-162°C and for oil ASV-5, within 174-176°C. This temperature can also be found from the beginning of the second bend of the system's electrical conductivity vs temperature curve (see Fig. 5.4, a, curve 1).

The influence of the method and rate of cooling on grease properties can be illustrated by the data presented in Table 5.3 [40,56]. A rapid cooling of LioSt greases (methods 1 and 2) is unacceptable as this results in a relatively low-strength structure. A slow cooling of the LioSt melt in the oil from 210 to 50-60°C at a rate of 1-2 deg/min (method 3) or a combined (stepwise) cooling, first at a rate of 1.5-3 deg/min down to 160-165°C and then at a rapid rate of 20-25 deg/min (method 5) is very effective. An isothermal crystallization of the soap dispersion in oil at t_1 = 185-187°C in 30-40 min at a combined cooling is proposed in [56], whose author substantiated the choice of these t_1 values by the existence of a phase transition of LioSt in oil at 185°C. Other researchers [57] believe a temperature of about 150°C to be the optimum one for an isothermal crystallization of LioSt in petroleum oils. The authors of [58,63] concluded that a temperature of 100 or 130°C is the best temperature of an isothermal crystallization for LioSt in oil MVP. It should be noted, however, that the authors of [58] did not qualitatively characterize the 12-hydroxystearic acid used by them. At the same time the electron-microscope structure, presented by these authors, very much resembles in appearance the structure of a LiSt-based grease, which makes it reasonable to believe that the composition of the 12-HoSt used by them contained a significant amount of such acids as stearic and palmitic. These examples proves that the composition and properties of the dispersion medium should be taken into account in determining the optimum temperature of isothermal crystallization of lithium hydroxystearate. The isothermal crystallization of LioSt in various dispersion media was ascertained [51-55] to be most effective at a temperature for which the change in the electrical conductivity, i.e., the rate of crystallization of LioSt in oil, is the maximum. As noted above, this temperature can be determined from the second bend of the electrical conductivity vs. temperature curve. For LioSt-oil ASV-5 systems, t_1 is 174-176°C; for an 1 : 1 mixture of oils AU and I-50A, t_1 is close to 170°C; and for oil S-220 it is 160-162°C [53,55].

It should be pointed out that the selection of the optimum temperature of isothermal crystallization is not only governed by the composition and properties of the dispersion medium, but depends also on the soap anion. The author of [59] determined that the optimum temperature t_1, for lithium stearate is 100°C with a cooling delay for a time not less than 2-3 h. Under otherwise equal conditions this yields a grease featuring the minimum penetration, which corresponds to the maximum shear strength value [51-56,58-63] or, as called in some of these papers, shear strength of the structure. It was, however, shown in [58] that at a slow cooling of the LiSt melt in petroleum oil MVP in a cooking kettle the maximum shear strength is exhibited by a grease that was crystallized for 2 h at t_1, = 130°C. The authors of [20,60,61], studying this dependence under rapid cooling conditions (a thin layer on a pan), ascertained that the highest thickening effect of the soap (the maximum shear strength and the minimum oil separation under pressure) for LiSt-vaseline oil and LiSt-oil MVP systems is as well attained at t_1 = 130°C and at a holding time of 0.5 h. The difference between the systems consists only in the absolute value of the shear strength, which decreases with increasing polarity of the dispersion medium, filament-like fibres with a very high an-isometricity, promoting the formation of a strong system, were found [60] to form in such a method of cooling of an isotropic melt of lithium stearate in oil. Note that $t_1 = 130°C$ is close to the temperature of formation of a crystalline LiSt.

At a slow cooling of the lithium stearate melt in petroleum oils, as compared with a rapid cooling, less strong systems are formed and t_1 shifts towards lower temperatures, approximately to 110°C [20,61]. With increasing LiSt concentration in oil the differences in the properties of greases by different cooling methods show up more markedly (Fig. 5.5) [61].

Studying the effect of the isothermal crystallization at 90, 110, 130, and 150°C in polysiloxanes and in AU-type petroleum oils produced from Anastasyev-, Zhirnov-, and Balakhny-field petroleums, the authors of [62] concluded that at cooling of greases on a pan in a 4-5 mm thick layer (rapid cooling) the best results for silicones are obtained at $t_1 = 110$-$120°C$; for the AU oils produced from Zhirnov petroleum, at $t_1 = 90°C$; and for the AU oils from Anastasyev and Balakhny petroleums, at $t_1 = 110°C$.

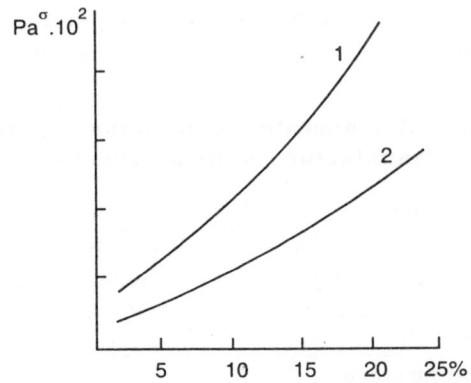

Fig. 5.5. Dependence of shear strength of lithium stearate-based greases on their cooling conditions [61]: 1-rapid colling; 2-slow cooling; Concentration of thickener, %; σ, Pa-10^2

Summarizing the results of studies on determining the optimum temperature of isothermal crystallization of lithium stearate in various oils [57-63], it can be stated that the temperature t_1 varies within 90-130°C depending on the dispersion medium composition and cooling method. It should be noted, however, that the papers having been cited do not indicate the purity of LiSt used by their authors, which should as well affect the optimum t_1 value. It seems that this temperature for LiSt dispersions in various dispersion media should be determined more exactly by determining the temperature dependence of the electrical conductivity of these systems and remembering that t_1 should correspond to its maximum value.

It is obvious at the same time that, in contrast to LioSt-based greases, for which a slow cooling and an isothermal crystallization at 160-180°C are preferable, it is expedient to cool LiSt-based greases rapidly, holding them for a certain time at 110-130°C. The grease holding time at t_1 should be 1-1.5 h in both cases.

Lithium greases can be produced with the use of SFA as the saponifiable stock [58]. The authors of [58] propose in this case to carry out the isothermal crystallization of soaps of commercial SFA as follows: for C10-C16 ones at $t_1 = 100°C$ (a high colloidal stability) or at $t_1 = 130°C$ (maximum shear strength and drop point, good viscous characteristics); for C17-C20 ones, at

$t_1 = 130°C$; and for a wider fraction, C10-C20, at $t_1 = 100°C$. The use of C10-C16 SFA yields the best results [58]. SFA-based lithium greases have, however, a poor quality and have found no practical application.

Sodium greases (Constalins) behave in cooling similarly to LiSt-based greases. They are usually cooled in a thin layer by feeding the soap melt in oil onto a water-cooled surface of a rotating drum, from which the grease is removed by pressed-on blades. An isothermal crystallization of sodium greases is not practiced.

Hydrated calcium greases in batch production method are usually cooled by feeding coolant (water) into the cooking kettle jacket and at the same time treating the soap concentrate with the remaining part (1/2-2/3 of the calculated amount) of oil. Fatty cup greases should be treated with a thin oil stream (slowly), while synthetic ones may be cooled more rapidly.

Anhydrous and calcium complex greases are not very sensitive to cooling conditions. Thus, the shear strength at 50°C and the colloidal stability of an anhydrous calcium grease at a cooling rate of 10 deg/min were 260 Pa and 25%, and at a rate of 0.5 deg/min, 280 Pa and 24% respectively [8, 37].

TABLE 5.4. Effect of temperature conditions of complex calcium grease manufacture on its properties

Temperature conditions (maximum thermal treatment temperature, 250°C)	Shear strength, Pa		Colloidal stability, %	
	day after manufacture	month after manufacture	day after manufacture	month after manufacture
Without isothermal treatment	350	1160	10.2	3.1
Isothermal treatment at heating stage (160-170°C, 20-25 min)	460	1120	8.8	3.1
Isotermal treatment (crystallization) at heating stage (160-170°C, 20-25 min) and at cooling stage (160-170°C, 35-40 min)	1900	1960	2.8	2.6

To improve the thickening effect of a calcium complex soap and to stabilize the grease properties with time, it is recommended to effect an isothermal crystallization of cCa-greases for 35-40 min at 160-170°C at the cooling stage [30]. As noted above, it is expedient to subject these greases to an isothermal treatment within the same temperature range also at the heating stage. The effect of such a production technique is illustrated by the data presented in Table 5.4 which were obtained for greases prepared on synthetic oils (polysiloxanes) and a calcium complex stearate and acetate soap. The expedience of an isothermal treatment of calcium complex greases of such a composition in a temperature range of 160-170°C is ascribed by the authors of [30] to the fact that the calcium stearate and acetate complex formation process begins at 130°C, proceeds at the highest rate at 150-160°C, and terminates at 170°C; they found these temperature transformations by means of a differential thermal analysis. Taking into account the process rate and termination, these authors proposed, with the aim to produce a stable grease, to carry out its isothermal treatment at 160-170°C both at the heating and at the cooling stage. However, the results for other dispersion media and calcium complex soap composition may be different; this should be taken into account in the practice of production of calcium complex greases.

Aluminium complex greases should be cooled rapidly, preferably with the use of scraper coolers. Thus, a cAl-grease produced with cooling at a rate of 10-12 deg/min exhibits a higher quality than that of a grease cooled at a rate of 2-4 deg/min (Table 5.5).

Table 5.5. Effect of cooling rate on properties of aluminium complex greases (thickener content, 11%)

Cooling rate, deg/min	Penetration at 25°C, 10^{-4} m	Drop point, °C	Shear strength, Pa		Viscosity at 0°C and 10 s^{-1}, Pa·s	Colloidal stability %
			20°C	80°C		
10-12	220	>250	1040	430	300	2.4
2-4	250	>250	560	300	180	4.7

A correct specification of the optimum conditions of the process of cooling of soap greases is very important as this provides not only a high quality product, but also savings in the stock and power consumption during production. The optimum cooling conditions, as also other process stages, are determined experimentally for every grease in its development and noted down in the production process schedule, which should be strictly observed.

Homogenisation (mechanical treatment) is the main stage in production of silica gel, bentone, pigment, and other similar greases as well as one of the important final steps in preparation of most soap and polyurea greases [1,2,4-6, 27, 64-75]. It usually precedes the packing and is often combined with the filtration and deaeration of greases. The homogenisation increases uniformity of the thickener distribution in the oil, thereby enhancing its thickening effect, provides a uniform texture of greases and improves their appearance. Furthermore, homogenisation results in the breakdown of condensation and formation of coagulatory thixotropic structure of a grease, improving its mechanical and colloidal stability and other characteristics. In the simplest case the homogenisation is effected by pressing the grease through a mesh or a system of screens, through narrow (30-50 mm) gaps of roller mills [4-6]. The homogenisation (e.g., of cup greases) by a single- or a multistage pumping of the grease by a rotary gear or a screw pump is sometimes regarded as quite adequate. Such homogenisation methods, however, involve low strain rates (up to several thousand inverse seconds), which yields poor results. Greases homogenised at low strain rates exhibit such shear strength, viscosity and colloidal stability, which under an intense mechanical action in the course of operation undergo significant changes and after subsequent idleness of the mechanism do not return to their initial values [66]. More perfect machines, which have gained a wide industrial application for homogenisation of greases, are valve-slot homogenizers, operating at high pressures and rotary homogenizers (colloid mills), operating at low pressures [1,4-6, 27,65-75], which are primarily used in the final stage of grease production for a single homogenisation. At the same time a repeated homogenisation at every production stage is successfully used in continuous processes [5, 27]; in this case the product at a relatively low pressure differential circulates through homogenising slot valves installed directly in the process piping.

The homogenisation of soap greases can result either in an increase or in a decrease in the shear strength and viscosity. If the homogenisation has not fully destroyed the condensation structure of the grease, then its shear strength and viscosity decrease after the homogenisation. Note that the condensation structure of a grease is almost fully conserved during cooling of an isotropic soap melt in an oil in the state of rest; this is particularly typical for lithium stearate-based greases. When the product is stirred during cooling and then the grease is pumped, the bonds of its condensation structure get partly and sometimes fully broken down and do not get restored later on. In case of a preliminary destruction of the condensation structure and a nonuniform distribution (dispersion) of a soap in an oil the homogenisation upgrades the rheological characteristics of greases. The ultimate result of the homogenisation of greases consists of two effects: the effect of breakdown of remaining bonds of the condensation structure and the effect of a more uniform and fine dispersion of the thickener in the dispersion medium, i.e., improvement of the thickening power of the soap. Depending on which of the effects is stronger, the values of rheological characteristics of the grease will either increase or decrease with respect to the initial characteristics of the nonhomogenised grease. It is essential that the grease properties after the homogenisation get stabilized, and undergo no significant changes in the course of service and under an intense mechanical action. Attaining this calls for a correct determination of technical parameters of the homogenisation. These parameters for rotary homogenizers (colloid mills) are the amount of the gap between the stator and the rotor, the rotor rotation speed and the flow rate of grease or the strain rate gradient and for valve-slot ones, the amount (height) of the gap between the valve and the seat and the grease pressure difference across the homogenizer, which as well determines the strain rate gradient. The homogenisation temperature is important in both cases. In the course of homogenisation the grease temperature rises by 5-15°C. Thus, during homogenisation of grease Litol-24 in a valve-slot homogenizer its temperature at a homogenisation pressure of 6 MPa rose by 6°C, and at 11 MPa, by 9.5°C [67]. The grease temperature rise in rotary homogenisers is greater than in valve ones [68].

Optimum homogenisation conditions are known for every grease type; they depend on the composition of the grease thickener: hydrocarbon, organic, inorganic or soap. Note that there exists a relation between the change of the properties of greases during the homogenisation process and their mechanical stability [65, 69, 71]. Special attention to the selection of the homogenizer type and homogenisation conditions should be given in production of silica gel, bentone, and soap greases; both the cation and the anion of the soap is of importance for the latter. Thus, for lithium hydroxystearate-based greases such as Litol-24 during homogenisation in rotary homogenizers the stability of their properties is attained at a strain rate gradient of $10^5 - 4 \times 10^5$ s^{-1}, and in a valve-type homogenizer, at a pressure of 24-25 MPa [68]. The authors of [68] believe those homogenisation conditions for lithium hydroxystearate greases to be the optimum and recommend the use of a valve-type homogenizer. The homogenisation of Uniol-type complex calcium greases changes their characteristics to a greater extent than those of lithium greases. Increasing the strain rate gradient in a rotary homogenizer to 10^5 s^{-1} steeply raises the shear strength of these greases; at its value of about 3.5×10^5 s^{-1} the shear strength stabilizes, increasing by a factor of 1.8-2.2 over the initial one (for nonhomogenisd grease) [68]. A similar change of the shear strength of calcium complex greases occurs during their homogenisation in a valve-type homogenizer at a pressure up to 24 MPa, but the grease viscosity in this case increases more considerably (by 40-70% over the initial value) than with the use of a rotary homogenizer (not over 15%), and therefore the authors of [68] prefer the use of colloid mills for the homogenisation of calcium complex greases. Table 5.6 presents recommendations on the use of homogenizers and the optimum homogenisation conditions for some grease types.

Table 5.6. **Homogenizers and homogenation conditions, recommended for some grease types**

Grease type	Homogenizer	Homogenization conditions		
		Temperature, °C	Pressure, MPa	Shear rate, s^{-1}
Lithium (such as Litol-24)	Valve-type	40-50	20-27	—
Sodium and sodium-calcium (such as 1-13)	as above	80-85	10-13	—
Complex calcium (such as Uniol)	Colloid mill	40-50	—	3.5×10^5-5.5×10^5
Complex aluminium (such as Alyumol)	Valve-type or colloid	40-50	10-13	3.5×10^5-5.5×10^5
Silica gel (such as Siol, Aerol)	Valve-type	20-70	20-27	—
Bentone	as above	20-70	20-27	—
Polyurea (such as Polimol)	Valve-type or colloid mill	40-50	25-30	at least 5×10^5

It should be noted that for many greases, such as those based on urea derivatives, on some organic and other thickeners, the homogenisation problems have been studied inadequately and the optimum conditions as well as the most efficient homogenizers have not been determined so far.

Deaeration. Many stages of the grease production process, such as the mixing of components in mixers and the homogenisation, especially in rotary homogenizers, involve ingress of air into the grease, which impairs its appearance, reduces its density, and degrades the chemical stability in storage [1,4-6,76,77]. To remove air from greases, they are deaerated in special units, deaerators, under a vacuum. The degree of the vacuum depends on the suction capacity of the pump used for the discharge of the deaerated grease. The grease deaeration process proceeds most efficiently when the vacuum acts on thin grease films having great specific surfaces. The deaeration rate rises with increasing vacuum, grease fragmentation degree (increasing specific surface of the films) and grease residence time in the working zone of the deaerator (Fig. 5.6). The grease deaeration degree depends also on the viscosity of the grease being deaerated [76]. Thus, under equal deaeration conditions (vacuum, 85%; t = 20 s; d = 0.5 mm; and t =25°C) and nearly equal contents of air in the grease (about 5 vol. %) the deaeration degrees at grease viscosities of 166 and 62 Pa·s (20°C and 10 s^{-1}) amounted respectively to 39 and 90%, i.e., 1.9 vol.% air in the former and 0.5 vol. % air in the latter case remained in the grease deaerated under these conditions. As can be seen from Fig. 5.6, the thickness of the film of the grease being deaerated (viscosity of 166 Pa.s at 20°C and 10 s^{-1}) should not exceed 0.3-0.5 mm. It should be taken into account, however, that as the width of gaps between plates decreases and the grease viscosity increases, the power required for forcing the grease through the slots rises. The thickness of the slots should be adjustable within 0.1-0.5 mm.

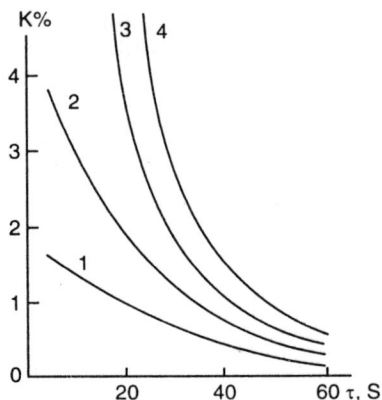

Fig. 5.6. Dependence of deaeration depth K on grease residence time in vacuum chamber (vacuum 85%) of deaerator and on grease film thickness-"δ", i.e., gap between plates [76].
δ = 0.3 mm (1); 0.5 mm (2); 1 mm (3); and 2 mm (4).

Filtration. Presence of abrasive mechanical impurities in greases substantially impairs their service properties, results in a premature wear of the elements being lubricated [1, 4-6, 77]. Moreover, presence of mechanical impurities and foreign inclusions significantly complicates the homogenisation of greases and often results in damages of homogenizers. In view of this, it is expedient to filter the starting components of a grease at the initial stage of the production process. Greases are also filtered at final stages of their production by forcing them through metallic screens or through self-cleaning disc or plate filters. The efficiency of screen filters is, however, very low.

The filtration of greases in continuous "closed" production processes at a thorough preparation (filtration) of the starting components and accomplishment of the whole process in a closed cycle (circuit) is generally not needed.

REFERENCES

1. Boner K.D. ; Production and Application of Greases. Transl. from English, Ed. by Sinitsyn V.V., Gostoptekhizdat, Moscow, 1958.
2. Velikovsky D.S. ; Poddubny V.N., Vainshtok V.V. and Gotovkin B.D., Consistent Greases, Khimiya, Moscow, 1966.
3. Ischuk Yu. L. ; Trends in Development of Process Layouts in Production of Plastic Lubricating Greases, in: Plastic Greases, Proc. Of Sci. and Techn. Conf. (Berdyansk, 1971), Naukova Dumka, Kiev, 1971, pp. 6-11.
4. Vainshtok V.V. ; Obelchenko E.I., Chernyavsky A.A. and Yashin-skaya M.S., Improvement of Plastic Lubricating Grease Production Processes, TslTIITEneftekhim, Moscow, 1978.
5. Ishchuk Yu.L. ; Fuks I.G. and Froishteter G.B., Production of Plastic Lubricating Greases, in: Diagram Manual of Production Layouts of Petroleum and Gas Processing, Ed. by Bondarenko B.N., Khimiya, Moscow, 198^, Ch. 2, pp. 97-104.
6. Bakaleinikov M.B. ; Plastic Lubricating Grease Production ׅ ׅ ׅ ׅ ־minder List For Operator), Khimiya, Moscow, 1977.

7. Fuks I.G., Plastic Lubricating Greases, Khimiya, Moscow, 1972.
8. Ishchuk Yu.L., Kuzmichev S.P., Krasnokutskaya M.E. et al., The Present State and Future Prospects for Advances in Production and Application of Anhydrous and Complex Calcium Greases, TsNIITEneftekhim, Moscow, 1980.
9. Proishteter G.B. ; Trilisky K.K., Ishchuk Yu.L. and Stupak P.K., Fiheologic and Thermophysical Properties of Plastic Lubricating Greases, Khimiya, Moscow, 1980.
10. Oparina E.M. ; On Problem of Plastic Lubricant Production Technology, in: Improvement of Grease Production Technology, GOSINTI, Moscow, 1961, pp. 12-20.
11. Ermilov A.S. and Tsvetkova K.I. ; Production of Consistent Greases by Intermittent and Continuous Methods with Use of Ultrasound, Ibid., pp. 21-31.
12. Rudakova N.Ya. ; Modern Design Features in Production of Consistent Greases, Ibid., pp. 32-51.
13. Garzanov G.E. ; Plan of Conversion of Existing Grease Production Plants and Factories, Ibid., pp. 64-80.
14. Nakonechnaya M.B. ; Ishchuk Yu.L., Goshko N.S. and Guberman B.N., Production Technology and Basic Scheme of Continuous Production of Complex Calcium Greases, in: Production and Improvement of Quality of Plastic Lubricating Greases, TsKIITEneftekhim, Moscow, 1970, pp. 45-53.
15. Ishchuk Yu.L. ; Composition, Structure, and Properties of Lithium Oxystearate-Thickened Greases, Cand. Tech. Sci. Thesis, Moscow, 1973.
16. Ishchuk Yu.L. ; Investigation into Influence of Dispersed Phase on Structure, Properties, and Technology of Plastic Lubricating Greases, Dr. Tech. Sci. Thesis, Moscow, 1978.
17. Nakonechnaya M.B. ; Complex Calcium Greases, Their Composition, Preparation, Structure, and Properties, Cand. Tech. Sci. Thesis, Moscow, 1970.
18. Svishevskaya G.I. ; Investigation into Properties of Plastic Lithium-Calcium Greases, Cand. Tech. Sci. Thesis, Moscow, 1971.
19. Puks I.G. ; Additives to Plastic Lubricating Greases, Khimiya, Moscow, 1982.
20. Shchegolev G.G. ; Microstructure and Properties of Lithium Greases, Cand. Chem. Sci. Thesis, Moscow, 1968.
21. Sinitsyn V.V. ; Ishchuk Yu.L. and. Prokopchuk V.A., Influence of Free Alkalis and Acids on Structure and Properties of Hydrated Calcium Greases, Dokl. AIT SSSR, Vol. 167, No. 2, pp. 1143-1145.
22. Sinitsyn V.V. ; Ishchuk Yu.L. and Ishchuk L.P., Free Alkalis and Acids in Plastic Lubricating Greases, Khimiya i Tekhnologiya Topliv i Masel, 1972, No. 9. pp. 23-26.
23. Ishchuk Yu.L. ; Influence of Dispersed Phase Composition on Structure and Properties of Oleogels, Plastic Lubricating Greases, in: Physico-Chemical Mechanics and Lyophilicity of Disperse Systems, Naukova Dumka, Kiev, 1983. Issue 15, pp. 63-75.
24. Ishchuk Yu.L. ; Sinitsyn V.V. and Krentkovskaya O.Ya., Surfactants in 12-Oxystearic Acid-Based Complex Calcium Greases, Khimiya i Tekhnologiya Topliv i Masel, 1972, No. 4, pp. 22-25.
25. Metering Pumps and Units. Catalogue of VNIIGidromash, TsINTI-khimneftemash, Moscow, 1975.
26. Ishchuk Yu.L. ; Maksimilian A.P., Mankovskaya U.K. and Losovaya V.I., Effect of Formulation and Production-Process Factors on Production of Complex Aluminium Greases, Khimiya i Tekhnologiya Topliv i Masel, 1985, No. 9, pp. 24-26.
27. Yurtin L.0. ; Tyshkevich L.V., Chmil V.S. and Shevchenko V.L., Experimental Study of Continuous Process of Production of Lithium 12-Oxystearate-Based Greases, in: Plastic Greases, Abstr.

28. Karpov N.F. ; Improvement of Production Technology of Natural Fat-Based Hydrated Calcium Greases, Neftepererabotka i Neftekhimiya, 1978, No. 8, pp. 14-16.
29. Krasnokutskaya M.E. ; Takonechnaya M.B. and Kuzmichev S.P., Influence of Heating Rate and Cooling Conditions on Properties of Complex Calcium Greases, in: Chemistry and Technology of Lubricants, Naukova Dumka, Kiev, 1977, pp. 75-79.
30. Krakhmalev S.I. ; Klimov K.I., Pavlov I.A. et al., Determination of Required Stages of Thermal Treatment of Complex Calcium Greases, in: Plastic Lubricating Greases, Moscow, 1977, pp. 92-96 (Proc. VNIINP, Issue 24).
31. Krentkovskaya O.Ya., Ishchuk Yu.P. and Sinitsyn V.V. ; Effect of Maximum Temperature of Preparation on Properties of cCa-Greases Thickened with 12-Oxystearic Acid Soaps, Neftepererabotka i Neftekhimiya, Kiev, 1971, Issue 4, pp. 107-109.
32. Ishchuk Yu.L. ; Novoded R.D., Zhebrovskaya N.V. and Klimen-ko P.L., Effect of Maximum Temperature of Preparation and of Thermal Treatment Time on Thermooxidation Stability of Complex Calcium Greases, Ibid., 1980, Issue 18, pp. 41-44.
33. Suchanek V. and Liebl X. ; Tixotropie plastickych masiv na basi Ca-acetatoveho komplexu, Ropa a Uhlie, 1968, Vol. 10, No. 12, pp. 659-665.
34. Major G. ; Mozes G., Fabian G. et al., Erfahrungen auf dem Gebiet der Theorie. Herstellung und Verwendung von Calcium-Komplex-Schmierfetten, Schmierstof. und Schmierungstechn., 1967, Nr. 21, s. 31-43.
35. Pat. 2,822,331 (USA). Anhydrous Calcium 12-Hydroxy Stearate Grease. J.P.Diluorth, O.P. Purycar, H.V. Ashburn. Publ. : Febr. 4, 1958
36. Pat. 2,862,884 (USA). Process for Anhydrous Calcium 12-Hydroxy Stearate and Estolide Containing Grease. J.P. Dilworth, C.H. Culane, R.F. Kelson. Publ. Dec. 2, 1958.
37. Dugina L.N. ; Nakonechnaya M.B. and Krasnokutskaya M.E., Temperature Conditions of Production of Anhydrous Calcium Greases, in: Plastic Greases, Abstr. of Reports at 3rd All-USSR Sci. and Techn. Conf. (Berdyansk, Sept., 1979), Naukova Dumka, Kiev, 1979, pp. 112-114.
38. Krasnokutskaya M.E., Ishchuk Yu.L., Nakonechnaya M.B. et al, Structurization and Phase Transitions in Soap-Oil Systems, in: Physico-Chemical Mechanics of Disperse Systems and Materials, Naukova Dumka, Kiev, 1983, Pt. 1, pp. 175-176.
39. El-Shaban Ibragim Muchamed Husein, Study of Influence of Production-Process Factors on Efficiency of Action of Additives in Lithium Greases, Cand. Tech. Sci. Thesis, Moscow, 1981.
40. Lendyel I.V., Cherednichenko G.I., Shkolnikov V.M. et al., Effect of Physico-Chemical Characteristics of Dispersion Medium on Low-Temperature Properties of Plastic Lubricating Greases, in: Neftepererabotka i Neftekhimiya, Naukova Dumka, Kiev, 1974, Issue 14, pp. 35-39.
41. Lozovaya V.I., Dagaev V.A. and Mankovskaya N.K., Dependence of Properties of Pseudogels of Aluminium Soaps in Oil on Temperature Conditions of Their Production, in: Plastic Greases, Proc. of 2nd All-USSR Sci. and Techn. Conf. (Berdyansk, 1975), Naukova Dumka, Kiev, 1975, pp. 52-55.
42. Ishchuk Yu.L., Mankovskaya N.K., Maksimilian A.P. and Lendyel I.V., Complex Aluminium Greases Based on Synthetic Acids, in: Neftepererabotka i Neftekhimiya, Naukova Dumka, Kiev, 1975, Issue 13, pp. 12-16.
43. Cambell J.D. and Harting G.L., Lithium Complex Greases, Ind. Lubr. and Tribol., 1976, Vol. 28, No. 5, PP. 160-164.

44. Ishchuk L.P., Bulgak V.B., Ishchuk Yu.L. and Fedorei A.I., Effect of Additions of Adipic and Sebacic Acids on Thickening Power of Lithium 12-Oxystearate, Khimiya i Tekhnologiya Topliv i Masel, 1978, No. 2, pp. 26-28.
45. Bulgak V.B., Ishchuk Yu.L. and Godun B.A., Use of the DTA Method in Development of Technology of Complex Lithium Greases, in: Plastic Greases, Abstr. of Reports at 3rd All-USSR Sci. and Techn. Conf. (Berdyansk, Sept., 1979), Naukova Dumka, Kiev, 1979. pp. 203-204.
46. Bulgak V.B., Ishchuk L.P. , Godun B.A. and Klimenko P.L. , Determination of Temperature Conditions of Production of Complex Lithium Greases, Khimiya i Tekhnologiya Topliv i Masel, 1981, No. 4, pp. 14-15.
47. Bulgak V.B., Ishchuk Yu. L. and Godun B.A., Investigation into formation of Complex Soaps and Rheologic Characteristics of Greases Based on Them, in: Physico-Chemical Mechanics of Disperse Systems and Materials, Naukova Dumka, Kiev, 1983, Pt. 1, pp. 180-181.
48. Smietana M., Smary kompleksowe, Wiad. naft., 1968, No. 10, pp. 225-228.
49. Dodon V., Jorga D. and Lima I., Nee unsori lubrificante, Petrol si Gase, 1968, Vol. 19, No. 5, pp. 314-316.
50. Inventor's Certificate 654,673 (USSR). Plastic Lubricating Grease. G.I.Svishevskaya, S.A.Stepanyants, A.V.Lotareva et al. Publ. in B.I., 1979, No. 12.
51. Podlennykh L.V., Ishchuk Yu.L. and Dagaev V.A., Study of Lithium 12-Oxystearate Crystallization from Melt in Vaseline Oil, Colloidn. Zhurn., 1979, Vol. 41, No. 2, pp. 369-371.
52. Podlennykh L.V., Dagaev V.A. and Ishchuk Yu.L., Study of Processes of Structurization of Lithium 12-Oxystearate in Vaseline Oil, Ibid., 1981, Vol. 43, No. 3, pp. 589-591.
53. Podlennykh L.V., Ishchuk Yu.L., Dagaev V.A. and Prokopchuk V.A., Effect of Chemical Nature of Oils on Crystallization of Lithium 12-Oxystearate, in: Plastic Greases, Abstr. of Reports at 3rd All-USSR Sci. and Techn. Conf. (Berdyansk, Sept., 1979), Naukova Dumka, Kiev, 1979, pp. 134-136.
54. Ishchuk Yu. L., Podlennykh L.V., Structurization in Soap-Oil Systems, in: Abstr. of Reports at 2nd All-USSJR School on Colloid Chem. of Petroleum and Petr. Products (Drogobych, 1981), TsNIITEneftekhim, Moscow, 1981, pp. 51-52.
55. Podlennykh L.V., Study of Lithium Oxystearate Structurization in Grease Preparation Process, Gand. Tech. Sci. Thesis, Moscow, 1982.
56. Dagaev V.L., Vdovichenko P.17. and Buyanova Z.I., .Effect of Dispersion Medium on Phase Transitions in Lithium 12-oxystearate, Neftepererabotka i Neftekhimiya, 1977, No. 4, pp. 14-16.
57. Matthews J.B., The Structure of Lubricating Grease, J. Inst. Petrol., 1953, Vol. 39, No. 333, pp. 265-275.
58. Vainshtok V.V., Kartinin B.N., Karakash S.I. and Avchina S.L., Investigation of Lithium Greases Thickened with Soaps of Natural and Synthetic Acids, in: Consistent Greases, Gostoptekhizdat, Moscow, 1960, pp. 11-26 (Proc. of I.M.Gubkin MNKh i GP, Issue 32).
59. Void M.J., Colloidal Structure in Lithium Stearate Greases, J. Phys. Ghem., 1956, Vol. 60, No. 4, pp. 439-442.
60. Shchegolev G.G., Trapeznifcov A.A. and Astakhov I.I., Colloid-Chemical Properties and Micro structure of Lithium Consistent Greasee, Khimiya i Tekhnologiya Topliv i Masel, 1965, No. 8, pp. 48-55.
61. Shchegolev G.G., TrapezniJsov A.A. and Taranenco V.G., The Role of Thickener Concentration and Surfactant Addition in Soap-Oil Systems, Kolloidn. Zhurn., 1973, Vol. 35, No. 2, pp. 322-327.

62. Vainshtok V.V., Smirnova N.S., Levento .R.A. and ICartinin B.N., Influence of Cooling Conditions on Structure and Properties of Lithium Greases, in: Plastic Greases, Proc. of Sci. and Techn. Conf. (Berdyansk, 1971), Naukova Dumka, Kiev, 1971, pp. 63-67.

63. Vainshtok V.V., Isothermal Crystallization as Processing Technique in Pro-duction of Lithium Consistent Greases, in: Improvement of Grease Production Techno-logy, GOSINTI, Moscow, 1961, pp. 81-95.

64. Melnik D.T. and Lendyel I.V., Study of Effect of Homogenation on Low-Temperature Properties of Greases, in: Chemistry, Technology, and Application of Lubricants, Naukova Dumka, Kiev, 1979, pp. 87-92.

65. Vinogradov G.V., Mamakov A.A. and Pavlov V.P., Homogenation and Rheologic Properties of Plastic (Consistent) Greases, Izv. Vuzov, Neft i Gaz, 1960, Ho. 3, pp. 81-88.

66. Vladimirskaya P.A., Smiotanko E.A. and Ermilov A.S., Homogenation of Plastic Lubricating Greases in Apparatus of Various Types, in: Plastic Lubricating Greases, TsNIITEneftekhim, Moscow, 1977, PP. 105-109 (Proc. of VNIINP, Issue 24).

67. Melnik D.T., Froishteter G.B. and Ish.ch.uk Yu.L. , Homogenation of Lithium Greases in Valve-Type Homogenizers, Kbimiya i Tekhnologiya Topliv i Masel, 1974, No. 1, pp. 42-46.

68. Melnik D.T. and Froishteter G.B., Study of Homogenation Conditions for Lithium and Complex Calcium Greases, in: Plastic Lubricating Greases, Additives, and Cutting Fluids, Vses. Obyed. "Neftekhim", Moscow, 1975, Issue 5, pp. 285-290.

69. Klimov E.I. and Leontyev B.I., A New Method for Studying Mechanical Properties of Thixotropy of Consistent Greases under Uniaxial Tension Conditions, Khimiya i Tekhnologiya Topliv i Masel, 1958, No. 4, pp. 66-72.

70. Meshchaninov S.M., Interrelation between Laboratory Test Methods and Performance of Synthetic Consistent Greases, Proizv. Smazochn. Materialov, 1961, Issue 6/8, pp. 98-118.

71. Klimov K.I., Leontyev B.I. and Sinitsyn V.V., On Effect of Strain Rate on Bulk Mechanical Properties of Plastic Lubricating Greases, Kolloidn. Zhurn., 1964, Vol. 26, No. 2, pp. 200-206.

72. Romanyutin A.A., Prusak A.G., Maksimilian A.P. et al., Effect of Homogenation Conditions on Properties of Complex Aluminium Grease, in: Plastic Greases, Abstr. of Reports at 3rd All-USSR Sci. and Techn. Conf. (Berdyansk, Sept., 1979), Naukova Dumka, Kiev, 1979, p. 151.

73. Romanyutin A.A. , Shevchenko V.L., Prusak A.G. and Maksimili-an A.P., Properties of Complex Aluminium Grease at Various Homogenation Conditions, in Improvement of Quality of Lubricants and Efficiency of Their Application, TsNIITEneftekhim, Moscow, 1980, pp. 39-43.

74. Melnik D.T., lyutenko N.E., Matrosov I.V. and Svishevskaya G.I., Continuous Cooling and Homogenation of Sodium-Calcium Greases, in: Plastic Greases, Abstr. of Reports at 3rd All-USSR Sci. and Techn. Gonf. (Berdyansk, Sept., 1975), Naukova Dumka, Kiev, 1979, pp. 119-121.

75. Melnik D.T., Investigation of Optimum Methods of Homogenation of Plastic Lubricating Greases of Various Nature and Their Introduction into Production Practice, Ibid., pp. 105-106.

76. Yurtin L.0., Froishteter G.B., Lendyel I.V. and Shevchenko V.L., Investigation of Optimum Conditions of Deaeration of Plastic Lubricating Greases in Continuous-Action Unit, in: Plastic Greases, Proc. of 2nd All-USSR Sci. and Techn. Conf. (Berdyansk, 1975), Naukova Dumka, Kiev, 1975, pp. 48-50.

77. Froishteter G.B., Modern Production Equipment for Plastic Lubricating Grease Plants, in: Plastic Greases, Proc. of Sci. and Techn. Conf. (Berdyansk, 1971), Naukova Dumka, Kiev, 1971, pp. 11-14.

Chapter 6

Processing Conditions and Layouts of Grease Manufacutring Plants, Determination of Their Capacities

Numerous processing layouts for manufacture of greases by intermittent(batch), semicontinuous and continuous methods have been described in literature [1-27], but many of them have not been tested properly and therefore have not been implemented so far in the commercial production of greases. Only the processing layouts that have either already been operated at grease factories or can be successfully employed in the designing, of updating the existing grease production facilities or constructing new high-efficiency facilities will be discussed here.

It should be pointed out that, in contrast to other petrochemical processes, grease manufacture is a waste-free production. If a substandard product has been obtained, it is directed for reprocessing. Grease manufacture (except very rare events) does not involve any harmful gas emissions and sewage, and therefore requires no special purification or treatment plants for them.

6.1. EFFICIENT LAYOUT OF BUILDINGS, ARRANGEMENT OF EQUIPMENT, CONFIGURATIONS OF MAIN PROCESS AND AUXILIARY FLOWS OF STOCK, CONTAINERS, AND FINISHED PRODUCTS

The arrangement of equipment and disposition of its items in processing layouts of commercial grease manufacture depend on specific conditions of the arrangement of production facilities. The most efficient and widely used indoor facilities is a multi-storeyed arrangement of the equipment with their cascadewise disposition from the top downwards in the direction of the main process flow: the top floor accommodates intermediate reservoirs for a several-day reserve of the stock and reagents; the middle floor houses proportioning devices, reactors (autoclaves-contactors, cookers), evaporators, heat exchangers-heaters; the bottom floor accomodates, pumps, coolers, and finishing equipment: homogenizers, deaerators and filters. Accumulating tanks are arranged either in the bottom floor or at the same level as the filling room.

Auxiliary procedures (preparation of some components and delivery containers, filling and storage) are carried out in rooms adjoining the main building of the processing plant on two sides (Fig. 6.1).

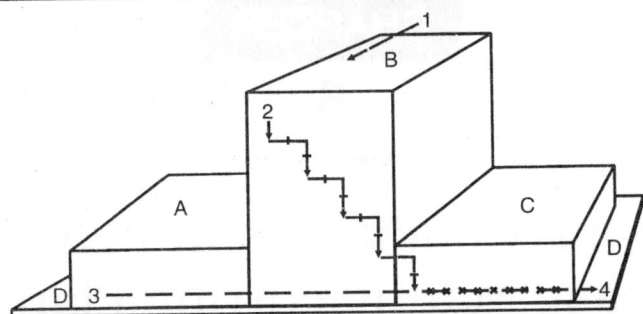

Fig. 6.1. Schematic layout of production rooms, main process flow, and auxiliary flows of stock, delivery containers, and finished products: A-storages of containers and reagents; B-processing plant building; C-filling area and finished product storages; D-loading/ /unloading ramp; 1-stock to plant; 2-main process flow; 3-containers to filling of products; 4-finished products to loading.

With such a layout of rooms and arrangement of equipment in them the main process flow (vertical) does not intersect auxiliary flows of the stock and delivery containers (side horizontal ones). This provides for the conveyance of the stock, containers, auxiliary materials, products and finished grease by the shortest distance, which results in high productivity of labour and in minimum power consumption.

Specific properties of greases and of intermediate products in their manufacture predetermine a close arrangement of the equipment units, as near the other as possible, the minimum length of pipings with the least number of bends and without excessive shut-off valves; pumps (first of all screw and gear ones) serving for the transfer of intermediate products and greases should be as close as possible to units from which a product is pumped out. Processing plants should occupy not over 20% of the area and volume of production rooms, while the remaining 80% are preferably allocated for intermediate storages of delivery containers and reagents, filling compartment and storage of finished products. The number and volume of accumulating tanks should be such as to exclude the work in filling rooms and storerooms in the evening and night time.

The most versatile processes, allowing a changeover from one grease type to another without appreciable difficulties, are intermittent ones. Continuous grease manufacture methods are highly effective, highly productive and economically advantageous, but only when the output of such a plant is at least ten thousand tons per year. At a lower output it is expedient to resort to semi-continuous manufacturing methods, where continuity of the process begins from either the thermal treatment or the cooling of the grease. For a small output (within 1000 t per year) it is advantageous to use the intermittent production method. These recommendations are, however, of a general character, and the selection of the manufacturing method should be thoroughly evaluated in every specific case. The main process layouts of grease manufacture will be discussed below. Conventional representations of the equipment in the flow diagrams are as follows:

Horizontal vessel with elliptical end plates

Vertical vessel with elliptical bottom and removable flat cover

Damping vessel

Vertical vessel with elliptical bottom, removable flat cover and jacket

Vertical vessel with conical bottom and removable flat cover and jacket

Vertical vessel with conical bottom and removable flat cover

Apparatus with turbine-type stirrer, with elliptical bottom, elliptical cover and jacket

Apparatus with propeller-type stirrer, elliptical bottom, removable flat cover and jacket

Apparatus with paddle-type stirrer, elliptical bottom, elliptical cover and jacket

Apparatus with scraper-and-paddle stirrer, with elliptical bottom, elliptical cover and jacket

Autoclave-contactor

Drying chamber

Buffer tank

Evaporator AV-600

Tube-in-tube heat exchanger

Shell-and-tube heat exchanger

Scraper-type heat exchanger

Electric-heating thermal unit

Flow mixer GART

Separator

Mixer-reactor with vortex layer of ferromagnetic particles

⌄	Drum-type cooler
►⊕►	Stationary filter
►⊕►	Filter in piping
►⊙►	Deaerator
►▭►	Valve-type homogenizer
⊘	Weighing batcher
⊡	Volume batcher
⊗	Volume-portion batcher
—⊙—	Pump
—⊘—	Metering pump
—⊕—	Vacuum pump
—⊚—	Air blower
Ⓜ⊘⊘⊘	Metering pump unit
⋈	Homogenizing valve

6.2. PLANT FOR MANUFACTURE OF SOAP GREASES BY INTERMITTENT METHOD

The flow diagram of manufacture of soap greases by the intermittent method is shown in Fig. 6.2. The plant provides for the preparation of greases with any soap, the latter being produced directly in the course of grease preparation, (in situ) by saponification of fats with metal hydroxides in 1/3-2/3 of the full amount of the dispersion medium. The plant can also be used to produce hydrocarbon greases. In this case scraper-type cooler 6 can be, if required, replaced with a drum-type cooler. Most hydrocarbon greases are not subjected to homogenisation. The grease cooling is often completed in reactor 4 by feeding a coolant into its jacket. This reactor is the basic unit in the intermittent grease manufacturing equipment. It should be fitted with a scraper-and-paddle stirrer. The following steps are effected successively in the reactor: mixing of components; saponification of fats or neutralization of acids; dewatering of the soap-oil concentrate; thermal treatment; and partial or full cooling by treating the soap melt with the remaining part of oil and feeding the coolant into the reactor jacket. The procedure of grease preparation with this layout is as follows. Reactor 4, with its central stirrer running, is charged by batchers 11 or by other proportioning methods (for example, reactor 4 can be installed on a strain gauge device and then the ingredients are fed by gravity) from vessels 1, 2, and 3 respectively with 1/3-2/3 of the calculated amount of the dispersion medium, the specified mass of saponifiable components and the corresponding amount of an aqueous solution (suspension) of the metal hydroxide. Then the heat carrier is admitted into the jacket of reactor 4 and the mixture is heated to 90-110°C with both stirrers running and the reactor content being circulated by pump 12 through homogenizing valve 13; saponification of fats takes place. When saponification has been completed (which is checked from the free alkali content), moisture is removed from the soap oil suspension, the completeness of the removal depending on the type of the grease being manufactured. At this stage the reactor is connected to vacuum pump 15 through condenser 14. After the moisture removal and an appropriate thermal treatment the soap oil concentrate is cooled, by treating it with the remaining part of oil as well as by feeding the coolant into the reactor jacket or by pumping the product after its treatment with oil through scraper type coolers 6. If the grease is not subjected to homogenisation, it is pumped either directly into delivery containers or into accumulating vessel 10 and then to the filling station.

Hydrated Ca-greases (cup greases) are manufactured by a simplified process, which is effected primarily in reactor 4, its jacket being fed during heating with steam and during cooling with water. When high temperature greases, such as lithium or calcium complex, are manufactured, the soap oil concentrate is after the de-watering heated to the required thermal treatment temperature (210-230°C) under continuous stirring and the soap melt having formed is held under these conditions for a time specified in the process schedule (10-40 min). Next, not stopping the stirring, the soap melt is cooled to 165-180°C by feeding the remaining amount of oil into the reactor and the coolant into its jacket. The soap melt cooling temperature is determined by the grease type and composition. The grease is held for the specified time at the temperature which is optimum for its type (isothermal crystallization) and, after an additional cooling to 150-160°C, additives from apparatus 5 are charged by metering pump 11 into reactor 4. If the additives are not thermally stable at this temperature, then they are fed to the second stage of scraper-type cooler 6, where the grease temperature at the inlet is 90-100°C.

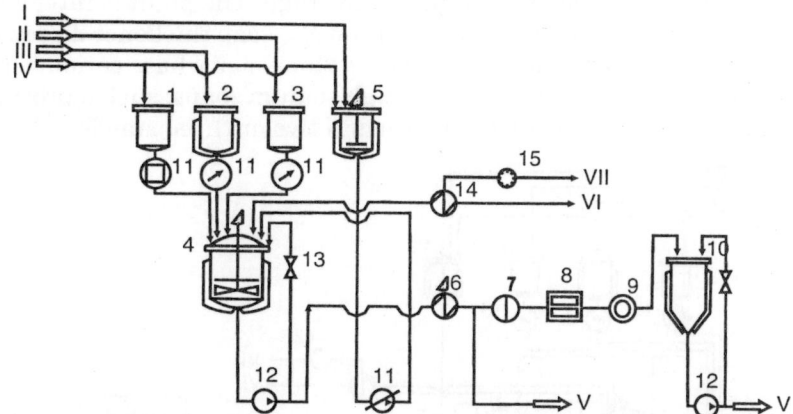

Fig. 6.2. Flow diagram of the plant for manufacturing soap greases by the batch method.
1-3-raw materials vessels; 4-scraper-and-paddle apparatus; 5-apparatus for a preparation of additives' concentrate; 6-scraper-type cooler; 7-filter; 8-homogenizer; 9-deaerator; 10-accumulating tank; 11-metering devices; 12-pumps; 13-homogenizing valves; 14-condenser; 15-vacuum pump; I-additives; II-saponifiable components; III-solution (suspension) of metal hydroxide; IV-oil; V-finished product (grease) for filling; VI-condensate; VII-air-uncondensed gases.

In scraper-type cooler 6, whose jacket is fed with a coolant, the grease is cooled down to 50-60°C. Used as the coolant is water, cooled to 3-5°C, which circulates in a closed circuit: scraper cooler is a Pre-refrigerated water cooler. If a refrigeration plant is not available, then demineralized water with a temperature of 20-23°C is used. Such a cooling is admissible only on condition of a thorough and deep treatment of water so that it would not contaminate the cooling surface of the cooler. At a severe contamination of the surface the treatment will turn out to be economically less advantageous than construction of a closed cooling system. The use of brine with a temperature of minus 10-15°C as the coolant in a closed cooling system is inexpedient as this steeply increases the viscosity of the product in the layer adjoining the wall, increases the power consumption by the drive, and eventually impairs the cooling conditions because of a great liberation of the dissipated heat. After the cooling, the grease passes in turn through filter 7, homogenizer 8, and deaerator 9, and then into accumulating tanks 10, from which it is filled into delivery containers.

In the intermittent production method it is possible to partly shorten the cycle and to raise the process efficiency by means of a concurrent feed of the reagents by metering pumps through a special mixer as well as by shortening the dewatering time by admitting additional heat through a heat exchanger incorporated into the circulation system of reactor 4.

With the modern equipment used, the intermittent process offers versatility, making it possible to manufacture practically any soap and hydrocarbon grease. The latter are produced in the reactor after dewatering of solid hydrocarbons (paraffin, ceresin, or petrolatum) at 105-110 °C and their dispersion in oil, followed by cooling in a scraper-type cooler or with the aid of a special cooling drum. The above described process layout is recommended for relatively small volumes (up to 2000 t per year) of grease manufacture.

6.3. PLANT FOR MANUFACTURE OF SOAP GREASES BY INTERMITTENT METHOD WITH USE OF AUTOCLAVE-CONTACTOR

An improved layout of manufacture of soap thickened greases (when natural fats or glycerides are used) by the intermittent method is shown in Fig. 6.3. It is distinguished by the use of autoclave-contactor 4 at the fat saponification stage. The plant is intended to manufacture various soap greases. Along with the preparation of a soap thickener directly in the grease manufacture process, it is also possible to prepare a thickener whose cation is a heavy metal, such as lead, by a double exchange reaction through sodium soaps; such a process is called the preparation of greases by the intermittent method in two or three stages.

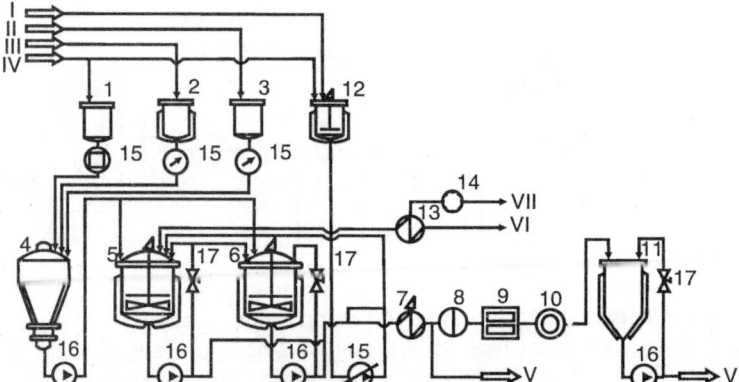

Fig. 6.3. Flow diagram of the plant for manufacturing soap greases by the batch method using autoclave-contactor. 1-3-raw materials vessels; 4-autoclave-contactor; 5,6-scraper-and-paddle apparatus; 7-scraper-type cooler; 5-apparatus for a preparation of additives' concentrate; 8-filter; 9-homogenizer; 10-deaerator; 11-accumulating tank; 12-apparatus for a preparation of additives' concentrate; 13-condenser; 14-vacuum pump; 15-metering devices; 16-pumps; 17-homogenizing valves; I-additives; II-solution (suspension) of metal hydroxide; III-saponifiable components; IV-oil; V-finished product (grease) for filling; VI-condensate; VII-air-uncondensed gases.

The process is accomplished as follows. Stock components of greases (dispersion medium, molten fats, aqueous solution or suspension of metal hydroxide) from vessels 1, 2 and 3 are in the specified proportions fed through batchers 15 into autoclave-contactor 4 which operates at an excess pressure of upto 1 MPa, where, at an elevated temperature (130-200°C, depending on the grease type), saponification of fats with formation of the soap-oil concentrate occurs in 20-40 min. The contactor is heated by a heat carrier circulating through its jacket. A hot reaction mixture under pressure from the contactor arrives into reactors 5 and 6, each fitted with a scraper and paddle stirrer, which operate in parallel (when required, in series). In the reactors the soap base is treated in hot condition with the second part of oil. Reactors 5 and 6 are provided with a system for removal and condensation of water vapour: a condenser 13 and a vacuum pump 14.

For most soap greases after the boiling off of water and thermomechanical dispersion of the thickener in reactors 5 and 6 (cycle duration, 2-3 h) the melt formed is cooled directly in the same reactors by the last amount of oil and then in scraper-type cooler 7. Additives and fillers from apparatus 12 are, depending on their function, composition and properties, fed in the form of solution or suspension in oil by metering pump 13 into reactor 5 or 6, or to the second stage of scraper-type cooler 7. The prepared grease is filtered (8), homogenized (9), deaerated (10) and then passes into accumulating tanks 11 for filling.

In the manufacture of greases thickened with soaps of polyvalent metals, Na-soap is as a rule prepared in contactor 4, from which the required soap is produced in reactor 5 by interaction with the corresponding water soluble salt of a heavy metal, and then water soluble salts are thoroughly washed away from the soap. Next, the moist soap is mixed, with oil and a part of the suspension is pumped out into reactor 6 to facilitate water removal. After the dewatering, the thermomechanical dispersion of the thickener in the oil is carried out in reactors 5 and 6. After an additional treatment of the melt with oil it is cooled down, the grease undergoes finishing operations and is filled into delivery containers.

At domestic factories, in contrast to the process layouts used abroad, the employment of an autoclave-contactor in the manufacture of greases has not found wide application. However, the incorporation of the contactor in a grease production plant where natural fats (glycerides) are employed as the saponifiable stock significantly shortens the grease preparation time and saves costly saponifiable components.

An intermittent process of manufacture of lubricating greases with the use of an autoclave-contactor of a special design, described in [25] deserves attention. The flow diagram of this process is shown in Fig. 6.4. The plant is intended to manufacture various soap greases. Autoclave-contactor 4 is charged with a part of the dispersion medium, saponifiable components and metal hydroxide, after which the temperature is raised to the specified level by feeding the heat carrier into the jacket and internal coil of the contactor.

The completion of saponification requires not over 30 min. When required, moisture can be removed from autoclave-contactor 4 by connecting it to a vacuum system (condenser 10 and vacuum pump 11). The grease preparation can be completed in contactor 4 by charging into it the remaining amount of oil and all the required additives and starting the feed of coolant into its jacket and internal coil. As the temperature of the contactor content decreases and its viscosity increases, a special-design high-speed stirrer mills (homogenizes) the circulating greases; this exerts a particularly favorable effect on the quality of lithium greases. In most cases, however, the soap concentrate from autoclave-contactor 4 is transferred by pump 12 into apparatus 5 with a scraper and paddle stirrer, into which the oil and additives are fed and where the grease preparation is completed. Next, the grease is fed through cooler 7 and homogenizer 8 to the filling station. As reported in 25, the cycle of preparation of a hydrated calcium grease in this plant lasts 1.5-2 h, and of a sodium or lithium grease, about 3 h.

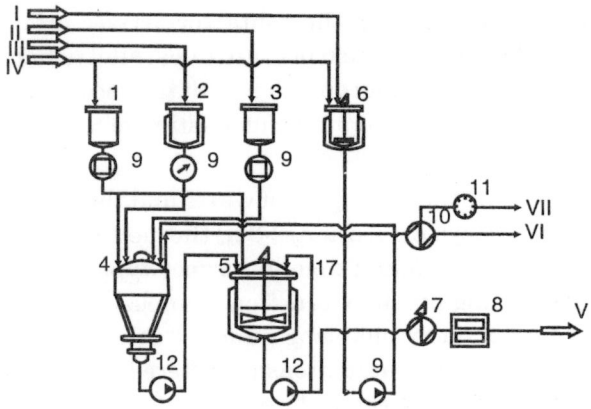

Fig. 6.4. Schematic diagram of "Stratco" plant for manufacture of soap greases by intermittent method; 1-3-stock vessels; 4-"Stratco" autoclave-contactor; 5-scraper-and-paddle reactor; 6-apparatus for preparation of additive concentrate; 7-"Votator" scraper-type cooler; 8-homogenizer; 9-metering devices; 10-condenser; 11-vacuum pump; 12-pumps; I-VII-the same as in Fig. 6.3.

6.4. PLANT FOR MANUFACTURE OF SOAP GREASES BY SEMICONTINUOUS METHOD

The advantages of both the intermittent and the continuous grease manufacturing process can be attained with the use of a semicontinuous process, which provides versatility and high productivity of the manufacture of greases as well as their optimum quality. The flow diagram of this method for a plant using the modern equipment and methods of in-process inspection of the quality of semifinished products and of the finished grease is shown in Fig. 6.5.

The plant is intended for manufacturing by the semicontinuous method any types of soap grease based on stearic, oleic, 12-hydroxystearic acid, natural and synthetic fats. Used as the dispersion medium can be oils of various nature (petroleum, synthetic and their mixtures) according to the field of application of greases and demands placed on them. The appropriately prepared stock from vessels 1-3, through batchers 16 is fed into reactors 4 fitted with high-speed stirrers that can intensely stir a low viscosity suspension. With good stability of the quality of the stock and reagents as well as with a high accuracy of their proportioning, such reactor-mixers can be substituted with more efficient mixing units, such as GART-type flow mixers.

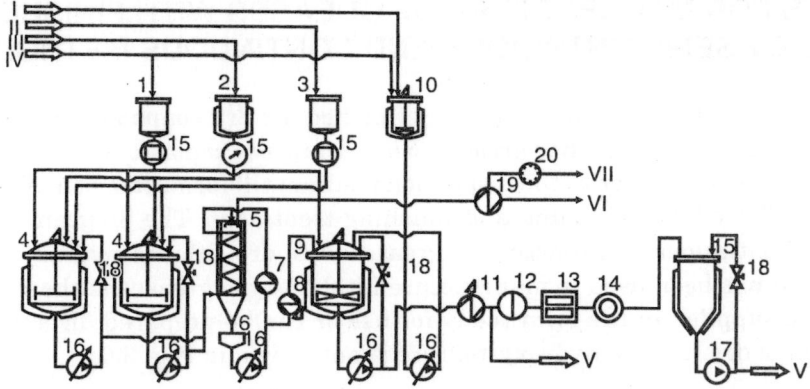

Fig. 6.5. Flow diagram of manufacture of soap greases by semicontinuous method: 1-3-stock vessels; 4-reactors-mixers; 5-evaporator; 6-buffer tank; 7-heat exchanger; 8-scraper-type heat exchanger; 9-scraper-and-paddle reactor; 10-apparatus for preparation of additive concent-rate; 11-scraper-type cooler; 12-filter; 13-homogenizer; 14-deaerator; 15-accumulating tank; 16-metering devices; 17-pump; 18-homogenizing valve; 19-condenser; 20-vacuum pump; I- VII-the same as in Fig. 6.3.

The saponified reaction mixture, prepared alternately in one of the reactors 4, which operate in parallel, is fed by metering pump 16 into a tubular coil of evaporator 5, where the mixture is dewatered under vacuum by means of the heat of the preheating in reactors 4 as well as the heat transferred by the heat carrier fed into the evaporator jacket. It is also possible to use additionally external heat exchanger 7, through which the product is circulated repeatedly till complete (if this is needed) moisture removal, the moisture content being monitored by a moisture meter. The dewatered mixture from the circulation loop is transferred by metering pump 16 through scraper-type heat exchanger 8 for the thermal treatment into reactor 9 equipped with a scraper-and-paddle stirrer, where the grease is held at the thermal treatment temperature for the time specified in the process chart. Next, with the stirrer running, the melt is treated with the remaining amount of oil. In this step, when lithium hydroxystearate greases are being prepared, the soap melt in oil is cooled to 160-180°C (feeding the coolant into the reactor jacket if required). The isothermal crystallization is carried out at the specified temperature. Next (if required), the grease is further cooled to 150-160°C and additives from apparatus 10 are charged into reactor 9 by metering pump 16, The concentrate of additives can also be fed to the second cooling stage of scraper-type cooler 11. When the additives have been mixed the contents of reactor 9 are transferred by pump 16 through scraper-type cooler 11, filter 12, homogenizer 13, and deaerator 14 into accumulating tank 15 and to the filling station.

The number of intermittently acting reactors 4 and their volume are selected so as to ensure at the specified output a continuous operation of the dewatering unit as well as of the whole train of equipment downstream of reactor 4. A long time experience gained in operation of semicontinuous plants for manufacture of general purpose greases confirms their attractiveness. The recommended output of such plants is from 2 to 10 thousand tons per year.

6.5. PLANTS FOR MANUFACTURE OF GREASES WITH FINISHED AIR-DRY SOAPS BY SEMICONTINUOUS AND CONTINUOUS METHODS

Commercial production processes using finished air dry soaps are employed in manufacture of sodium, lithium and other greases. Such a process consists of a thermomechanical dispersion of the soap thickener in the dispersion medium till formation of an isotropic (homogeneous) melt, followed by its cooling and finishing treatment. The diagram of an imported plant for a semicontinuous manufacture of greases with dry lithium hydroxystearate in the layout in which it has been used to manufacture Litol-24-type greases is shown in Fig. 6.6. A dry soap may be supplied in the form ready for use or can be prepared directly in the plant. The preparation of dry lithium hydroxystearate is not shown in Fig. 6.6.

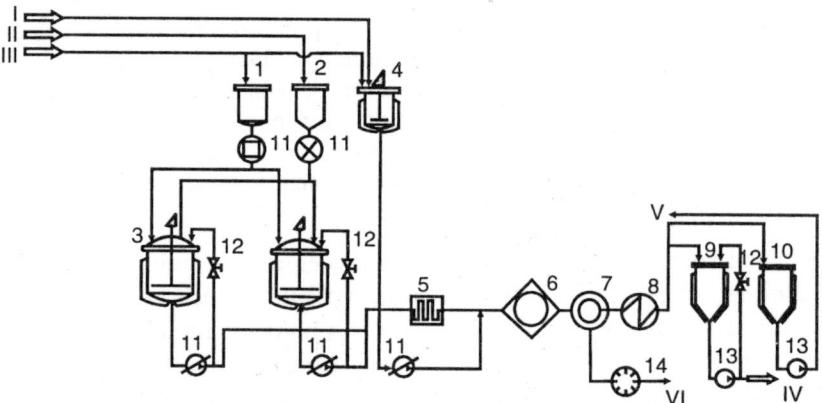

Fig. 6.6. Flow diagram of manufacture of greases with finished air-dry lithium oxystearate by continuous method: 1,2-stock vessels; 3-reactors; 4-apparatus for preparation of additive concentrate; 5-electric heater; 6-mixer; 7-deaerator; 8-scraper-type cooler; 9,10-accumulating tanks; 11-metering devices; 12-homogenizing valves; 13-pumps; 14-vacuum pump; I-additives; B-air-dry lithium oxy-stearate; ni -oil; IV-finished product (grease) for packaging; V-substandard product; VI-air.

To produce a dry soap, the saponifiable stock and an aqueous solution (suspension) of the alkali LiOH in the required amounts are mixed in alternately operating reactors, fitted with a stirrer and a jacket for the heat carrier feed. On the completion of the reaction of saponification or neutralization (for fatty acids) an aqueous suspension (slurry) of the soap arrives for drying into a continuous action vacuum drum type drier. The air dry soap is fed by an air-lift into vessel 2 and then by batcher 11 into one of two parallel-action reactors 3, where 2/3 of the required amount of oil is charged beforehand from vessel 1 through batcher 11. After a thorough stirring the suspension of soap in oil is pumped by metering pump 11 through tubular electric heater 5, where it is heated to 205-210°C. Next, the melt, mixing with the remaining amount of oil and with additives in mixer 6, arrives into deaerator 7; the solution (suspension) of additives in oil is fed into mixer 6 from apparatus 4 by pump 11. In deaerator 7 air is removed from the soapoil melt by the action of vacuum created by pump 14. Next, the melt is fed for cooling into scraper-type cooler 8. The cooled grease arrives into accumulating tank 9 and then to the filling station; a substandard product is through accumulating tank 10 directed for reprocessing. Note that reducing (homogenizing) valves 12 are installed in all circulation loops (the circulation is produced by pumps 11).

Apart from the above described layout, LZ-31 type greases are produced with the use of air-dry lithium stearate by a continuous method at domestic factories [7,10]. The essence of the processing layout (Fig. 6.7) is that molten stearic acid from vessel 1 and a heated aqueous solution of LiOH from vessel 2 are continuously pumped by metering pumps 3 through tubular heat exchanger 4 operating under a pressure of 2-2.3 MPa, where the soap, LiSt, is formed. The moist soap (suspension) passes into vessel 5 and then, through a distributor, into drying chamber 6, where it is dried in a stream of air, heated to 160-190°C, which is fed by high-pressure air blower 7 through air heater. The air-dry soap powder is collected in vessel 8, from which it is fed through batcher 9 into apparatus 10 and 11. At the same time other free-flowing components from vessel 12 through batcher 4 and the dispersion medium (oil) from vessel 13 through batcher 14 are charged into these apparatus.

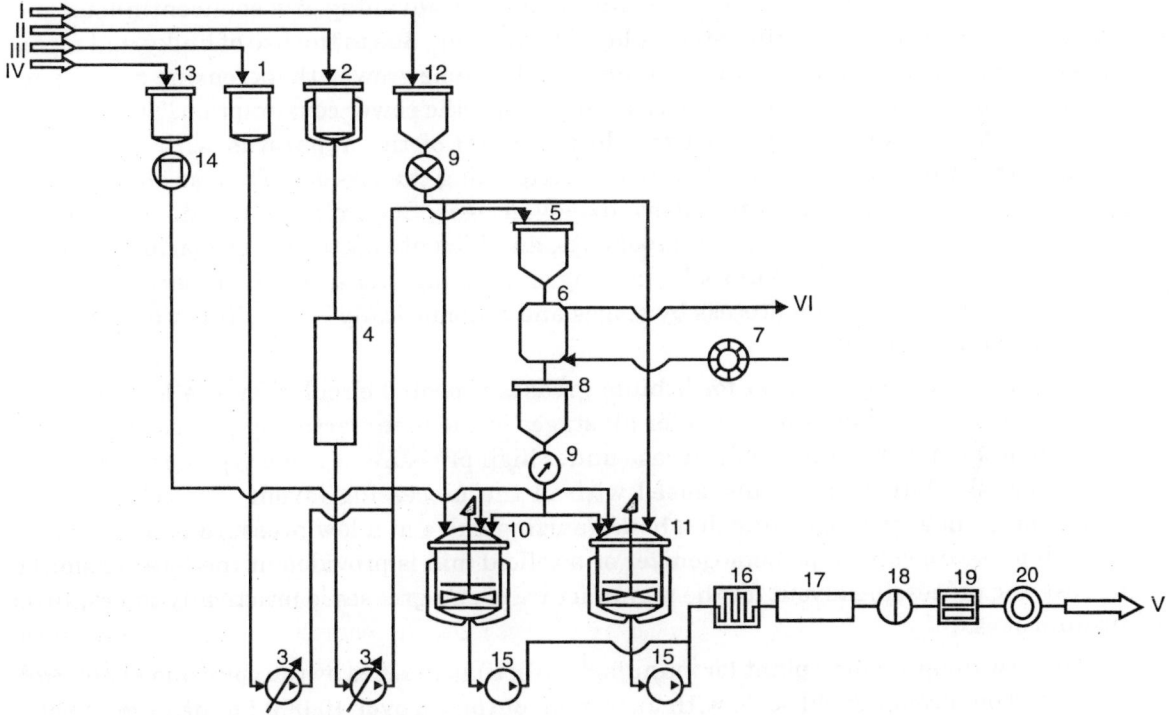

Fig. 6.7. Flow diagram of manufacture of greases with finished air-dry lithium stearate by continuous method: 1,2,5,8,12,13-stock and semiproduct vessels; 3-metering pumps; 4-tabular heat exchanger; 6-drying chamber; 7-air blower; 9,14-batchers; 10,11-scraper-and-paddle apparatus; 15-pumps; 16-thermal unit; 17-tubular cooler; 18-filter; 19-homogenizer; 20-deaerator; I-additives; II-lithium hydroxide solution (suspen-sion); III-commercial stearic acid; IV-oil; V-finished product (grease) for packaging; VI-used-up air.

Apparatus 10 and 11 operate alternately, their cycles being shifted with respect to each other. The suspension of soap in oil, containing also the required additives, is pumped by high-pressure pump 15 through thermal unit 16, where it is heated to 205-210°C, after which the melt arrives into tubular cooler 17, where it is cooled down to 50-60°C. The cooled grease is filtered 18, homogenized 19, and deaerated 20 and then filled into delivery containers. This plant has a low output, about 500 t/year. Grease manufacturing processes based on air-dry soaps turned out on the whole to be uneconomical and have found no wide application in the manufacture of general purpose greases. It is to be noted that the above described domestic plant 6.7 for manufacture of greases with the use of finished soaps is more efficient than an imported plant [16].

6.6. PLANTS FOR MANUFACTURE OF SOAP GREASES BY CONTINUOUS METHOD

At a plant output over 10,000 t/year, well-known disadvantages of semicontinuous and intermittent grease manufacturing process layouts show up, due to the use of bulky and metal intensive apparatus with stirrers, whose number and volume grow with increasing unit capacity of plants. This causes a significant increase in the specific power consumption. The need for a cyclic actuation (operation with a time-shifted cycle) of the apparatus with stirrers in semicontinuous processes creates difficulties in controlling the process. The use of a continuous processing layout eliminates the above disadvantages. Such plants include a system of continuous proportioning of starting components, an efficient reactor for preparing the soap base, special design scraper type units for heating and cooling over a broad temperature range and devices for monitoring the process by stages and automatically controlling it for strength and viscosity characteristics.

It was ascertained [23] that for lithium greases repeated circulation at a low pressure through homogenizing valves provided for all stages of the manufacturing process yields the same effect as a single treatment of a grease under high pressure in valve type homogenizers, and therefore the latter can be dispensed with in the processing layout. For cCa greases, however, the homogenisation through a homogenizing valve at a low pressure is inadequate, and a high pressure valve type homogenizer or a colloid mill is provided at the final stage. As to the rest, the processing layout for the manufacture of cCa greases is practically the same as for lithium ones.

The flow diagram for a plant for manufacture of soap greases by the continuous method, implemented on a commercial scale with an annual output of over 10,000 t of lithium greases such as Litol-24, is shown in Fig. 6.8. Oil, saponifiable components, and the solution (suspension) of metal hydroxide (LiOH) from vessels 1-3 are fed by metering pump unit 4 into flow mixer 5 where they are mixed vigorously. Oil before its arrival into flow mixer 5 is heated to the specified temperature in heat exchanger 6. In flow mixer 5 there occurs not only the mixing, but also a partial saponification (neutralization) of fatty components. The degree of saponification depends on the composition of the saponifiable fats, temperature, time of product residence in the mixer and mixing efficiency. Next, the mixture enters a tubular coil of evaporator 7 where, at an elevated temperature (140-180°C) and a pressure of about 1 MPa, the fat saponification reaction is completed. The pressure in the system is controlled by a correspond-ing valve as a function of the temperature. An automatic system monitors and controls the free alkali content in the product at the outlet from the evaporator coil. The water-oil suspension of the soap from the coil passes into the evaporation chamber of unit 7, where it is dewatered under a combined effect of atomization in the volume and a film flow on the surface of the unit. The pressure in the evaporator is maintained near atmospheric pressure by means of sucking the vapour from the evaporation chamber by vacuum pump 8 through condenser 9. The temperature of the dewatered product at the evaporator bottom is maintained within 170-200°C by varying the flow rate of the heat carrier fed into the evaporator jacket. The required depth of vacuum in the evaporator is provided by means of an analyzer that monitors the moisture content in the product at the outlet from the unit.

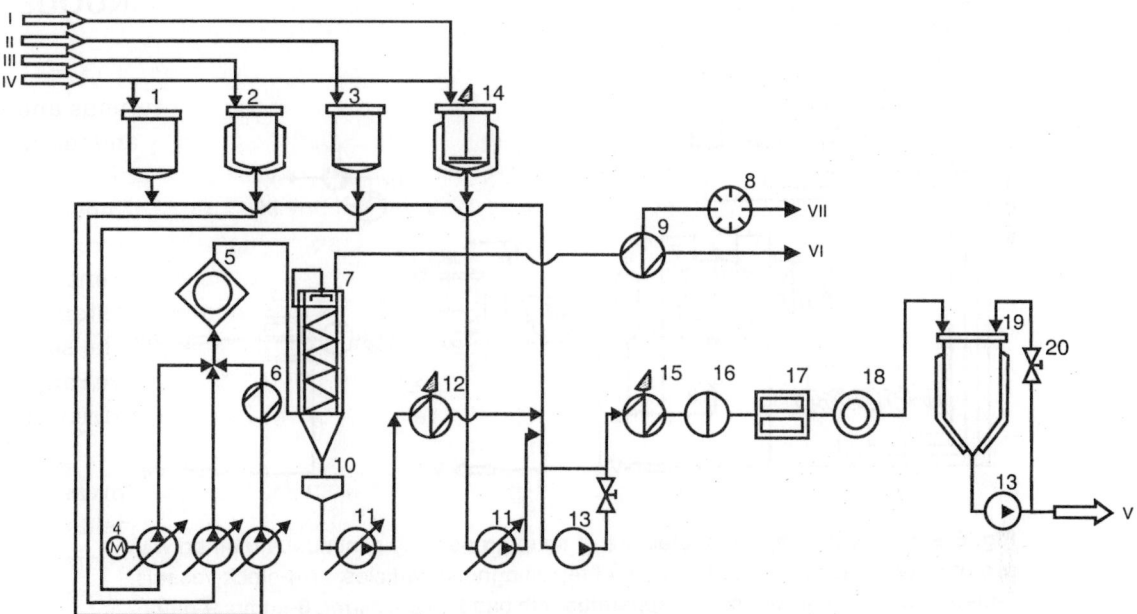

Fig. 6.8. Flow diagram of manufacture of soap greases by continuous method: 1-3-stock vessels; 4-metering pump unit; 5-mixer; 6-heat exchanger; 7-evaporator; 8-air pump; 9-condenser; 10-buffer tank; 11-metering pumps; 12-scraper-type heat exchanger; 13-pumps; 14-apparatus for preparation of additive concentrate; 15-accumulating tank; 20-homogenizing; 18-deaerator; II-Lithium hydroxide solution (suspension); III-commercial sraeric acid; IV-oil; V-finished product (grease) for packaging; VI-condensate VII-airs uncondensed gases.

The dewatered product is fed by metering pump 11, whose delivery is controlled from the level gauge of the evaporator, for the thermal treatment into scraper-type heat exchanger 12 and then, having mixed in pump 13 with the remaining amount of oil and with additives fed by metering pump 11 from apparatus 14, arrives into scraper-type cooler 15. A good mixing is attained by circulating the grease by pump 13 through homogenizing valve 20.

The cooled grease after scraper-type cooler 13 is filtered 16, homogenized 17, deaerated 18 and then fed into accumulating tank 19 and to the filling station. The grease quality -shear strength, viscosity, and penetration at the inlet to accumulating tank 19 is monitored by a special device which controls the feed of the remaining amount of oil and the grease cooling rate. A special accumulating tank is provided for a substandard product, from which the latter is returned by a metering pump for reprocessing.

The use of this continuous processing layout for the manufacture of general purpose greases yields the following significant advantages over the intermittent and semicontinuous processes: 5-6 times lower metal content of the basic equipment; 3-4 times smaller volume of production rooms; 25-30% lower capital outlays; 7-8% lower manufacturing cost of finished products, etc.

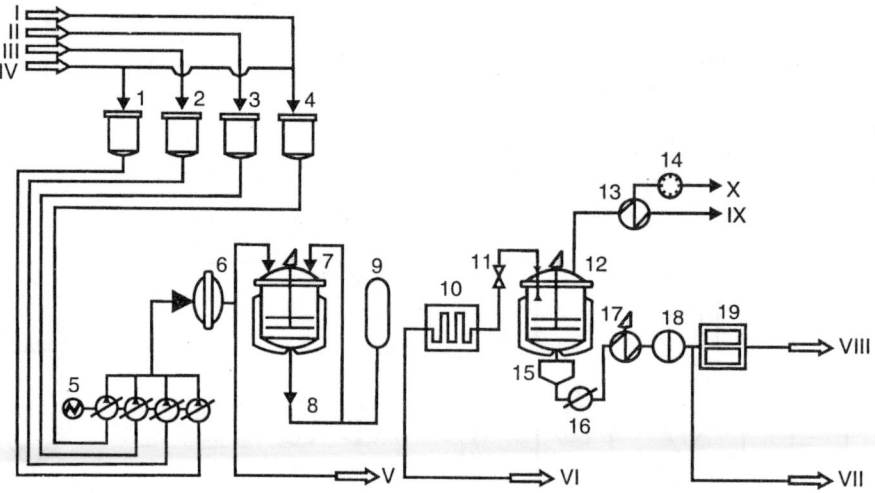

Fig. 6.9. Flow diagram of manufacture of soap greases by continuous method with sse of mixer-reactor with vortex layer of ferromagnetic particles: 1-4-stock vessels; 5-metering pump unit; 6-mixer; 7-apparatus with paddle-type stirrer; 8-pump; 9-damping vessel; 10-thermal unit; 11-pressure regulator; 12-evaporator; 13-condenser; 14-vacuum pump; 15-buffer tank; 16-metering pump; 17-scraper-type cooler; 18-filter; 19-homogenizer; I-additives; II-metal hydroxide solution (suspension); III-saponifiable components; IV-oil; V-VIII-finished product (grease) for packaging; IX-condensate; X-air and uncondensed gases.

Another continuous process, whose flow diagram is shown in Fig. 6.9 [18-22], has been implemented in the commercial production of lithium stearate greases (such as TsIATIM-201). Stock components from vessels 1-4 in the specified proportions are fed by metering pump unit 3 into mixer-reactor 6 with a vortex layer of ferromagnetic particles. The water-soap oil suspension having formed in the mixer arrives into apparatus 7, from which by high pressure pump 8 it is pumped through damping vessel 9, into thermal unit 10. The mixture here is heated to the temperature of an isotropic melt of soap in oil and then under a pressure of 2-2.5 MPa passes through a special throttling device into evaporator 12, where it is dewatered. The pressure in the evaporator is maintained close to atmospheric pressure by means of a vacuum system consisting of condenser 13 and vacuum pump 14. The dewatered grease is delivered by pump 16 through scraper-type cooler 17, where it is cooled to the specified temperature (50-60°C), passes through filter 18, and homogenizer 19 to the filling station.

The authors of [22] recommend to employ the plant under consideration as a multipurpose one for manufacturing lubricants of various types. Thus, synthetic cup grease or various cutting fluids are obtained at output V or VI, the remaining plant equipment being not used. When producing high temperature greases, sodium ones are obtained at output VII, and lithium and calcium complex ones, which require a thorough homogenisation, at output VIII.

An automated process control system provides for monitoring and automatically controlling the temperature, pressure, and level in all units of the plant in accordance with the values set on the control desk. The quality of the finished grease and intermediate products of its manufacture in this plant is monitored and controlled as well with the aid of special devices, described in Chapter 8.

6.7. PLANT FOR MANUFACTURE OF GREASES THICKENED WITH HIGHLY DISPERSED INORGANIC AND ORGANIC THICKENERS

The manufacture of greases based on inorganic thickeners (modified precipitated and pyrogenic silica gels, bentonite clays, pigments, carbon black, etc.) differs from the manufacture of soap greases and consists in a successive intense mechanical dispersion of the thickeners in oil with the use of various mixers and homogenizers.

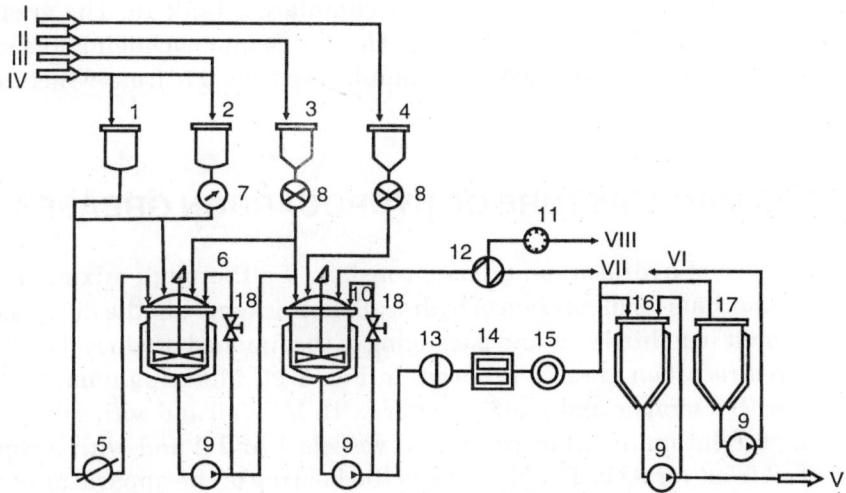

Fig. 6.10. Flow diagram of manufacture of organically and inorganically thickened greases: 1-4-stock vessels; 5-metering pump; 6,10-scraper-and-paddle apparatus; 7, 8-batchers; 9-pumps; 11-vacuum pump; 12-condenser; 13-filter; 14-homogenizer; 15-deaerator; 16,17-accumulating tanks; 18-homogenizing valves; I-graphite, molybdenum disulfide; II-aerosil, bentonite; III-additives; IV-oil; V-finished product (grease) for packaging; VI-substandard product; VII-condensate; VIII-air and uncondensed gases.

We will discuss some features of the manufacture of greases based on highly dispersed inorganic and organic thickeners by way of example of pyrogenic silica gel (aerosil). For greases based on precipitated silica gel the thickener is prepared directly in the plant, whereas the manufacture of greases based on pyrogenic silica gel employs a finished aerosil, modified with various surfactants. The flow diagram of manufacture of a Grafitol type silica gel grease is shown in Fig. 6.10.

The specified amount of the dispersion medium (oil) from vessel 1 is pumped by metering pump 5 into apparatus 6 with a scraper-and-paddle stirrer. Additives and, when required, also a dispersing agent from vessel 2 are charged through batcher 7. Next, the stirrer is turned on and the heat carrier feed into the jacket of apparatus 6 is started, after which the first part (80% of the calculated amount) of modified aerosil is charged from bin 3 through portioning batcher 8. To avoid lumping of the thickener in oil, it is charged by small portions during 7-8 h, the mass of the portion being progressively reduced from 5 to 1 kg. When aerosil is being charged into apparatus 6, pump 9 is turned on and the mixture is circulated through the loop: apparatus 6 - pump 9 - homogenizing valve 18 - apparatus 6. After a thorough uniform distribution of aerosil in oil has been attained, the mixture is transferred by pump 7 from apparatus 6 into scraper-and-paddle apparatus 10, into which graphite from vessel 4 is charged by batcher 8. The temperature of the mixture is raised to 105-110°C and the moisture that got into it with the starting components is removed by connecting apparatus 10 to vacuum pump 11 through condenser 12. Next, the remaining part (20%) of the thickener (aerosil) is charged into apparatus 10 in small portions. The mixture is stirred and circulated by pump 9 through homogenizing valve 18 to obtain a homogeneous mass, which is then filtered (13), additionally homogenized (14), deaerated (15), and transferred into accumulating tank 16. The grease from the latter, if it meets the specifications, is fed by pump 9 for filling and packaging. If a substandard product has been obtained, it is collected in accumulating tank 17, from which it is sent for reprocessing.

6.8. PLANT FOR MANUFACTURE OF HYDROCARBON GREASES

The manufacture of hydrocarbon greases consists of a thorough mixing of pre-filtered components, thermomechanical dispersion of hydrocarbon thickeners in the dispersion medium, rapid cooling of the melt in a thin layer and packaging of the finished product. The flow diagram of manufacture of hydrocarbon greases is shown in Fig. 6.11. Metering pump unit 4 charges apparatus 5, fitted with a scraper-and-paddle stirrer, with 1/3 of oil and with solid hydrocarbons (paraffin, ceresin, petrolatum, montan wax) from vessels 1 and 2 and with a concentrate of additives in 1/3 of oil from vessel 3. The stirrer and the heating of the apparatus are turned on, the temperature of its contents is raised to 120-140°C and at the same time the remaining part of oil is charged into it in a thin stream. Next, the melt is fed by pump 6 through filter 7 on to the surface of the rotating drum of cooler 8 and then to the filling station.

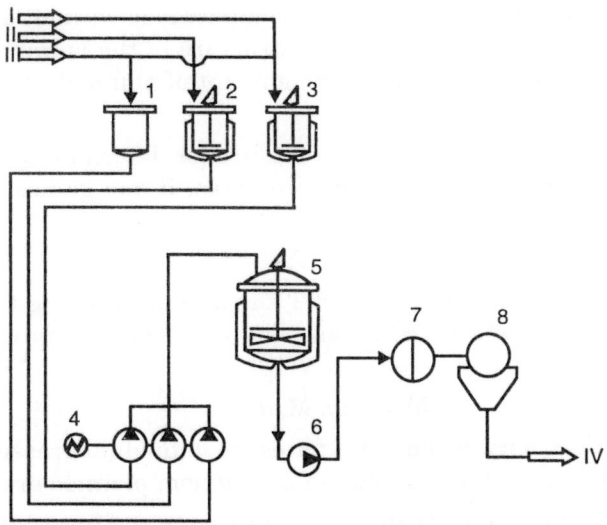

Fig. 6.11. Flow diagram of manufacture of hydrocarbon greases: 1-3-sleek vessels; 4-metering pump unit; 5-scraper-and-paddle apparatus; 6-pump; 7-filter; 8-drum-type cooler; I-additives; II-solid hydrocarbons; III-oil; IV-finished product (grease) for packaging.

6.9. DETERMINATION OF CAPACITY OF LUBRICATING GREASE MANUFACTURING PLANTS AND SHOPS

Due to existence of a wide variety of processing layouts for manufacture of a very broad assortment of greases, the capacity of individual plants and shops at production facilities is calculated for different manufacturing layouts in different ways: some determine it in tons of greases manufactured; others indicate the design capacity in terms of a certain grease type; and still others calculate it in tons of the "main" grease, etc. This prevents summing up the capacities of individual plants to determine the total capacity of a grease factory or of the grease industry as a whole.

A specific feature of the grease production, is that one and the same greases are manufactured by intermittent, semicontinuous, and continuous methods. The main units in intermittently and semi-continuously operating plants are scraper-and-paddle reactors or reactors-stirrers of other types, whose capacity can be of 1, 3, 5, 10, 16 m^3, etc. Several grease types with sharply differing manufacturing cycles, whose durations depend on the grease type, reactor volume, heating and cooling methods and other factors, are produced on one and the same plant by a semicontinuous and especially by an intermittent (batch) method. Different upper temperature limits for the manufacture of greases, e.g. up to 110°C for hydrated calcium greases, but up to 230°C for lithium complex ones, necessitate different methods of heating of apparatus, i.e., different plants can be used for certain groups (several types) of greases.

When the range of products is changed, the maximum amount of grease production, attainable on a plant during a year, changes as well, and the production capacity, if determined in tons of actually produced greases, on an yearly basis is incomparable. The degree of utilization of the design (planned) capacity as well cannot be determined without a special recalculation. Thus, any change in the range of products and structure in the manufacture of greases calls for recalculation of the capacity in terms of the amount of planned or actually manufactured products.

Considering the above, the following procedure for determining the grease production capacities is proposed [27]. With continuous processes, if one grease type is produced, the annual capacity is

$$C = \text{Out} \cdot T,$$

where Out is the hourly or daily output, and T the annual working time, hours or days. If several grease types are manufactured on one and the same continuous process plant, the annual production capacity is

$$M = \text{Sum } M_i \, a_i$$

where Mi is the annual capacity of the plant for every individual grease type and a_1 the fraction of every grease type in the total amount of production, planned or actual annual one.

The annual production capacity for semi-continuous and intermittent grease manufacturing processes should be determined separately for the following production types:
- with the maximum upper heating temperature up to 140 °C: hydrated and anhydrous calcium, hydrocarbon and other greases (type I);
- with the maximum upper heating temperature up to 240 °C: ordinary sodium and lithium, complex lithium, calcium and aluminium greases (type II);
- organically and inorganically thickened greases: carbon black, silica gel, bentone and others (type III).

The basic equipment to be used as the basis for calculating the capacity is a scraper-and-paddle reactor or a reactor of another type, where, depending on the grease type and plant type, the following procedures can be accomplished: mixing of components; saponification of fats; moisture removal; thermomechanical dispersion of the thickener; thermal treatment and a partial cooling of the soap melt. The output of this unit may be corrected depending on the throughput capacity of other manufacturing process stages (units): water evaporation (evaporator); thermal treatment (scraper-and-tube heaters); cooling (scraper-and-tube coolers); homogenisation (rotary or valve type homogenizers), etc.

To ensure the comparability of capacities of plants of the same type, they should be determined after reducing the amounts of production of various greases by the specified production types to tons of reference types of greases (products).

The annual capacity in semicontinuous and intermittent grease manufacture methods is

$$C = \frac{TY_c}{c} AK$$

where T is the annual working time of the equipment unit (apparatus), in hours h; Y is the yield of grease from the equipment unit (apparatus) per cycle (i.e., the output per revolution), tons; C is the cycle (revolution) time, h; A is the number of equipment units (apparatus), for which the capacity is calculated; and K is the factor of conversion of the output of greases to the reference product type.

If the plant includes equipment units (apparatus) of several types and sizes, then the capacity is

$$M = \Sigma \frac{T_i Y_{ci}}{\tau_{ci}} A_i K_i$$

where $\Sigma T_i Y_i A_i K_i, \tau_c$ – is the plant capacity in tons of the range of greases manufactured in it and Ki the factor of conversion of the output of the i-th grease to the reference product type.

The annual working time T is determined with allowance for the time of running repairs and major overhauls of the plants. The grease yield per cycle is determined on the basis of the data of the design or actual grease production process schedule. It is advisable to employ the Yc value actually attained in the period under consideration, which should be not below the planned value. The Yc value depends on the type and size of the apparatus and on the grease production technology. The annual working time and the grease yield per cycle are relatively constant quantities, while the cycle time τ_c varies depending on many factors, such as the stirring efficiency, heating rate (heat carrier type and temperature), stock quality, skill of the operating personnel (advanced work techniques), etc., and should therefore be revised and defined more exactly on a regular basis.

The conversion factor K is determined for every grease separately:

$$K = \frac{Y_{h,r}}{Y_h}$$

where $Y_{h \cdot r}$ is the yield of the reference grease (reference tons) per apparatus-hour, t, and Y_h, the yield of the grease manufactured in the plant per apparatus-hour, t :

$$Y_h = \frac{Y_c}{c}$$

The $Y_{h \cdot r}$ is calculated as the average value in the industry for apparatus volumes and production (plant) types (I, II, III).

The grease production structure, types, sizes and number of apparatus may change, and hence the average annual grease yield per apparatus-hour is not a constant value. Therefore, to determine it, one should have once in 5 years a complete calculation of the capacity in physical tons and data on the range of greases produced, types and sizes of apparatus, their number, actual cycle duration and product yield per cycle, and annual equipment working time. These data allow the calculation of the industry-average $Y_{h \cdot r}$, and then the capacity of plants by production types (I, II, III) can be determined in a simplified manner:

$$M = T_i A_i Y_{h \cdot r},$$

where T_i is the annual working time of the apparatus manufacturing the i-th grease, h; A_i is the number of the equipment units; and $Y_{h \cdot r}$ is the industry-average yield of the reference grease per apparatus-hour. The data on $Y_{h \cdot r}$ for the 11th five-year plan period for various production types and volumes of reactor units are presented in Table 6.1.

Table 6.1. Industry-average yield of reference grease, $Y_{h.r}$ t/apparatus-hour

Production (plant) type	Reaction unit volume, m^3			
	Up to 3	5	10	16
I	—	0.362	0.508	0.614
II	0.042	0.078	0.125	—
III	0.008	—	—	—

The capacity utilization degree is determined by dividing the amount of production of greases, converted into tons of the reference grease, by the capacity in tons of the same grease:

$$U = \frac{Q_1 K_i}{M} = \frac{Q_1(Y_{h.r}/Y_{hi})}{M}$$

where U is the capacity utilization factor, Q_1 the amount of pro-duction of individual grease types in plants of a given type, and $K_i = Y_{h.r}/Y_{h.i}$ the factor of conversion of the production of individual greases into the reference product type.

The above described procedure, a relatively simple one, makes it possible to determine the capacity of an individual factory, to compare the capacities, to evaluate the degree of their utilization and to plan their loading.

6.10. STAGE-BY-STAGE TEMPERATURE CONDITIONS OF MANUFACTURE OF MAIN TYPES OF SOAP GREASES

The temperature conditions of grease manufacture at various stages of the process may differ significantly depending on the soap anion and cation. The use of different fatty stock types for the manufacture of greases of one and the same type, such as hydrated calcium or ordinary lithium ones, often necessitates additional stages (steps) in the manufacturing process. The stage-by-stage temperature conditions for the manufacture of main soap grease types by the intermittent method in open apparatus are presented in Tables 6.2-6.9. It should, be noted, however, that other methods of manufacture of these greases are also existing. The tables generalize the grease manufacture conditions on the basis of either one and the same type of the thickener or the closeness of the upper temperature limits of their manufacture. It should be remembered that in some cases the thermal treatment and cooling rate, thermal treatment and isothermal crystallization times and homogenisation degree substantially affect the properties of some grease types.

Table 6.2. Temperature conditions (°C) of manufacture of hydrated and anhydrous calcium greases

Process stage	Synthetic cup grease °C	Fatty cup grease °C	Anhydrous calcium grease °C
Charging of dispersion medium (oil): for synthetic cup greases and anhydrous calcium greases, 50%; for fatty cup greases, 25-30% of calculated amount	30-60	30-60	30-60
Charging of SFA C_{20} and over (stillage residues) and SFA C_5-C_6 (C_5–C_9) in ratio of 3(2.5): 1	60-70	—	—
Charging of cottonseed oil and salomas in ratio of 3 : 1	—	50-70	—
Charging of 12-oxystearic acid	—	—	80-85
Charging of aqueous dispersion (40-45%) of $CaOH_2$ (lime milk or lime paste, or slaked lime suspension in oil) in amount for complete saponification (neutralization) of fats and provision of weakly alkaline (about 0.02% in terms of NaOH) reaction of finished grease	60-70	60-70	80-85
Heating and saponification of fats	95-100	95-100	95-100
Removal of excess water (dewatering)	105-110	110-115	105-110
Charging of water by portions (3-4 steps) or its feed by thin stream to ensure complete saponification of fats	—	110-115	—
Treatment of soap base (concentrate) with remaining part of oil	110-70	115-90	100-100
Thermal treatment and holding for 0.5-1 h	—	—	120-125
Cooling	60-65	60-65	80-85
Charging of additives	—	—	80-85
Final cooling, homogenization, and filling of grease into containers	60-65	60-65	60-65

Notes: 1. Process stages from neutralization of acids or saponification of fats to homogenization are effected under stirring.

2. Content of stillage residues and SFA $C_5 - C_7$ is of 12-14 %; of cottonseed oil and salomas %; and of 12-oxystearic acid, of 11-13 %.

Table 6.3. Temperature conditions (°C) of manufacture of calcium-sodium and sodium greases

Process stage	Grease 1-13 °C	Constain °C
Charging of dispersion medium (oil), 25-30% of calculated amount	30-60	30-60
Charging of castor oil	50-60	50-60
Charging of salomas	—	50-60
Charging of aqueous dispersion (40-45%) of time milk, or lime paste, or slaked lime suspension in oil in amount for obtaining 2-2.5% Ca-soap for finished grease	80-85	—
Charging of aqueous solution (35-40%) of NaOH in amount for complete saponification of fats and provision of weacly alkaline reaction (0.02% NaOH) of finished grease	80-85	80-85
Heating and saponification of fats	95-100	95-100
Dewatering of reactio mixture and temperature increase at rate of 2-3 deg/min	140-145	135-140
Charging of remaining part of oil (for grease 1-13, 30-35% of calculated amount) by thin stream	140-135	140-135
Thermal treatment	155-160	170-175
Charging by thin stream of remaining part (40-45%) of oil and additive Naftam-2	150-140	—
Repeated thermal treatment	155-160	—
Cooling (of Constain, rapid on cooling drum in thin layer or on pan) and filling of grease into containers	60-70	60-70

Notes:
1. Process stages from saponification of fats are effected under stirring.
2. At cooling of grease 1-13 in the cooker it is usually circulated in the cooking (kettle) – pump – cooking (kettle) loop.
3. The content of castor oil fats for grease 1-13 is of 19-21 mass %; of castor oil-salomas for Constain, of 18-22 mass %.

Table 6.4. Temperature conditions (°C) of manufacture of ordinary lithium greases

Process stage	With lithium Stearate °C	With lithium Oxystearate °C
Charging of dispersion medium (oil) : for greaseTsIATIM-201, 30-25%; for litol-24, 50% of calculated amount	30-60	30-60
Charging of stearic acid	80-85	—
Charging of 12-oxystearic acid	—	80-85
Charging of 10-11% aqueous solution of LIOH in amount for neutralization of acids and provision of weakly alkaline (0.02% NaOH) reaction of finished grease	85-90	85-90
Heating, neutralization of acids, correction of free alkali content, removal of water	100-130	100-120
Temperature increase	145-150	—
Charging by thin stream of remaining part (65-70%) of oil and of antioxidation additive	145-150	—
Temperature increase	150-210	120-210
Thermal treatment	205-210	205-210
Charging of remaining part of oil by thin stream and cooling	—	180-185
Isotermal crystallization and charging of antioxidation additive	—	180-185
Cooling (for grease TsIATIM-201, rapid: on pans or in cooling columns; for Lithol-24, in scraper-type coolers)	210-70	175-70
Filtration, homogenization, and filling into containers	70-60	70-60

Notes: 1. Process stages from neutralization of acids to filtration are effected under stirring.

2. Contents of stearic and 12-oxystearic acids are of 12-12.5% and 10-11% respectively.

Table 6.5. Temperature conditions (°C) of manufacture of complex lithium greases

Process stage	Temperature, °C
Charging of dispersion medium (oil): 65-70% of calculated amount	30-60
Charging of 12-oxystearic acid	80-85
Charging of 10-11% aqueous solution of LiOH in amount for neutrallization of 12-HoSt and of complexing components (boric and terephtalic acids) and provision of weakly alkaline (0.02% NaOH) reaction of finished grease	80-85
Heating and neutralization of 12-HoSt	95-100
Charging of 20% aqueous solution of boric acid	95-90
Neutralization of boric acid and dewatering of reaction mixture	95-105
Continuation of reaction mixture dewatering and temperature increase	130
Charging of terephtalic acid, its neutralization and formation of complex soap	13-140
Temperature increase, thermal treatment, and holding of reaction mixture at thermal treatment temperature for 20-30 min	235-240
Charging of remaining part (30-35%) of oil by thin stream	240-180
Cooling	180-150
Charging of additives	150-145
Continuation of cooling	145-60
Filtering, homogenization, and filling of grease into containers	60-50

Notes: 1. Process stages from neutralization of 12-HoSt to filtration are effected under stirring.

2. The content of 12-oxystearic, boric, and terephtalic acids at their approximate mass ratio of 4:2:1 is of 12-15 mass %.

Table 6.6. Temperature conditions (°C) of manufacture of complex calcium and complex aluminium greases

Process stage	Calcium grease °C	Aluminium grease °C
Charging of dispersion medium (oil): for calcium greases, 75%; for aluminium greases, 50% of calculated amount	30-60	30-60
Charging of 40% solution of monohydrolyzed aluminium isopropylate in oil (15-16% of calculated amount of oil for grease)	—	30-60
Heating of contents in apparatus	—	120
Charging of SFA C_{10}-C_{20} heated to 60-70°C	60-70	120-110
Production of Al-soap of SFA and partical distillation off of isopropyl alcohol	—	120-125
Charging of benzoic acid, production of complex Al-soap, and removal of isopropyl alcohol	—	120-125
Charging of slaked lime suspension in oil (20% of calculated amount of oil)	60-70	—
Charging of acetic acid	60-70	—
Heating of reaction mixture, neutralization of acids, correction of alkali, and partial removal of moisture	95-100	—
Heating of reaction mixture, complete removal of moisture and isopropyl alcohol	100-160	125-205
Thermal treatment	225-230	205-210
Charging of remaining (30-35%) part of oil and partial cooling of the contents of cooking unit	—	210-170
Cooling and charging of concentrate of additives in oil	230-160	170-160
Cooling, homogenization and filling of grease into containers	60-65	60-65

Notes: 1. Process stages from the heating of content in the apparatus to homogenization are effected under stirring.
2. The total content of SFA C_{19}-C_{20} and benzoic acid at their equilmolar ratio is for cAl-greases of 9-11 mass%, and for cCa-greases (SFA C_{10}-C_{20}: Hac = 1.5-2), of 11-15 mass %.

Table 6.7. Temperature conditions (°C) of manufacture of complex barium greases

Process stage	Temperature °C
Charging of dispersion medium (oil) in calculated amount	30-60
Charging of acids isolated from cottonseed oil	30-60
Charging of estolides of 12-oxystearic acid	30-60
Charging of SFA C_7 - C_9	30-60
Charging of acetic acid	30-60
Charging of barium oxide octohydrate in amount for saponification (neutralization) of fatty components and provision of weak acidic reaction (0.3-0.8 % HOL) of finished grease	30-60
Heating, saponification (neutralization) of fatty components, and correction of alkalinity (acidity)	95-102
Continuation of saponification (neutralization) of fatty components, evaporation of water, and correction of alkalinity (acidity)	100-115
Temperature increase	115-140
Thermal treatment	140-145
Cooling and check of grease drop point	140-100
Charging of antioxidation additive	100-95
Cooling, homogenization, and filling of grease into containers	95-60

Note: Process stages from charging of barium oxide octohydrate charging to homogenization are effected under stirring.

PROCESSING CONDITIONS AND LAYOUTS ...

Table 6.8. Temperature conditions (°C) of manufacture of aluminium greases

Process stage	Temperature °C
Charging of dispersion medium (oil): 30-35 % of calculated amount	30-60
Charging of moistened soap whose anion is formed by stearic and oleic acids	30-60
Temperature increase, water settling and draining	60-70
Heating and holding of soap concentrate	95-100
Dewatering	100-150
Charging of remaining part (65-70 %) of oil	150-130
Temperature increase and thermal treatment	200-205
Cooling and filling of grease into containers	60-70

Notes: 1. Process stages from the temperature increase to 60-70°C are effected under stirring.
2. The content of stearic and oleic acids at their ratio of 2 : 1 is of 19-21 mass %.
3. The moistened soap is prepared by double-exchange reaction.

Table 6.9. Temperature conditions (°C) of manufacture of polyurea greases

Process stage	Polyurea formation	Finished Polyurea	Polyureaacetate complex
Charging of dispersion medium (oil) : 30-35% of calculated amount	30-50	30-50	30-50
Charging of oil suspension of diisocyanate (10-15% of calculated amount of dispersion medium)	30-50	—	30-50
Temperature increase	60-80	—	60-80
Charging of oil suspension of amines (10-15% of calculated amount of dispersion medium)	60-80	—	60-80
Charging of oil suspension of finished polyurea (20-30% of calculated amount of dispersion medium)	—	60-80	—
Vigorous stirring (mixing) and temperature increase	180-200	90-110	120-130
Charging of oil suspension of fatty acids and calcium acetate (remaining (35-50%) part of dispersion medium	—	—	120-130
Charging of remaining (35-50%) part of dispersion medium under stirring	180-200	90-110	—
Vigorous stirring and homogenization at 5×10^5 s^{-1}	—	90-110	—
Temperature increase and thermal treatment	180-200	200-210	190-200
Cooling, filtration, homogenization at 5×10^5 s^{-1}, deaeration, and filling of grease into containers	40-60	40-60	40-60

REFERENCES

1. Boner K.D., Production and Application of Greases. Transl. from English, Ed. by Sinitsyn V.V., Gostoptekhizdat, Moscow, 1958.
2. Rudakova K. Ya., Modern Design Features in Production of Consistent Greases, in: Improvement of Grease Production Technology, GOSINTI, Moscow, 1961, pp. 32-51.
3. Garzanov G.E., Plan of Conversion of Existing Grease Produc-tion Plants and Factories, Ibid., pp. 64-80.
4. Nofsinger C.W. and Walas S.W., Grease-Manufacturing Methods Have Been Improving, Oil and Gas J., 1962, Vol. 60, No. 6, pp. 138-142.
5. Velikovsky D.S., Poddubny V.K., Vainshtok V.V. and Gotovkin B.D., Consistent Greases, Khimiya, Moscow, 1966.
6. Green W.B., Texaco Continuous Grease Manufacturing Process, KLGI Spokesman, 1969, Vol. 32, No. 10, pp. 368-373.
7. Afanasyev I.D., Vorobyeva V.A., Dobkin I.E. et al., Method of Continuous Production of Lithium Greases with Dry Soap, in: Pro-duction and Improvement of Quality of Plastic Lubricating Greases, TsKIITEneftekhim, Moscow, 1970, pp. 35-44.
8. Nakonechnaya M.B., Ishchuk Yu. L., Goshko N.S. and Guberman B.N., Complex Calcium Grease Manufacturing Technology and Layout, Ibid., pp. 45-50.
9. Smiotanko E.A., Peskov V.D., Vainshtok V.V. et al., Continu-ous Manufacture of Plastic Lubricating Greases, in: Plastic Grea-ses, Proc. of Sci. and Techn. Conf. (Berdyansk, 1971), Naukova Dumka, Kiev, 1971, pp. 69-71.
10. Dobkin I.E., Meshchaninov S.M., Zabelina N.P. et al., Experience Gained in Operation of Plant for Continuous Manufacture of Lithium Greases with Dry Soaps, Ibid., pp. 72-73.
11. Rosenzweig M.D., Compact Units are First to Make Grease Con-tinuously, Chem. Eng., 1971. No. 9, May 3, pp. 67-69.
12. Shekhter Yu. N., Vainshtok V.V., Puks I.G. et al., Universal Plant for Continuous Production of Consistent Greases, in: Production and Improvement of Quality of Plastic Lubricating Greases, TsNIITEneftekhim, 1970, pp. 25-35.
13. Ishchuk Yu. L., Trends in Development of Process Layouts in Production of Plastic Lubricating Greases, in: Plastic Greases, Proc. of Sci. and Techn. Conf. (Berdyansk, 1971), Naukova Dumka, Kiev, 1971, PP. 6-11.
14. Fuks I.G., Plastic Lubricating Greases, Khimiya, Moscow, 1972.
15. Ishchuk Yu. L., Composition, Structure, and Properties of Lithium Oxystearate-Thickened Greases, Cand. Tech. Sci. Thesis, Moscow, 1973.
16. Bronfin I.B., Belousov A.N., Leoshkevich E.A. and Slepchenko L.G., Experience Gained in Comissioning of Lithium Grease Complex at Omsk Integrated Petroleum Processing Works, in: Plastic Greases, Proc. of 2nd All-USSR Sci. and Techn. Conf. (Berdyansk, 1975), Naukova Dumka, Kiev, 1975, pp. 24-25.
17. Smiotanko E.A., Ermilov A.S., Krakhmalev S.I. and Pavlov I..V., Development of Technology of Manufacture of Plastic Lubricating Greases on Pilot Plant in Electric and Acoustic Fields, Ibid., pp. 53-54.
18. Obelchenko E.I., Yashchinskaya M.S., Chernyavsky A.A. and Vainshtok V.V., Continuous Process of Production of Lubricants with. Use of Electromagnetic Mixers, Nefteperabotka i Neftekhimiya, 1977, No. 7, pp. 10-12.

19. Kuzmichev S.P., Ishchuk Yu. L., Stepanyants S.A. et al., Experience Gained in Operation of Commercial Semicontinuous-Action Plant for Manufacture of Uniol-Type Greases, Khimiya i Tekhnologiya Topliv i Masel, 1977, No. 3, pp. 19-22.
20. Bakaleinikov M.B., Plastic Lubricating Grease Production Plant (Reminder List for Operator), Khimiya, Moscow, 1977.
21. Ishchuk Yu. L., Investigation into Influence of Dispersed Phase on Structure, Properties, and Technology of Plastic Lubri-cating Greases, Dr. Tech. Sci. Thesis, Moscow, 1978.
22. Vainshtok V.V., Obelchenko E.I., Chernyavsky A.A. and Yashchinskaya M.S., Improvement of Pistic Lubricating Grease Manufacturing Processes, TsNIITEneftekhim, Moscow, 1978.
23. Yurtin L.O., Tyshkevich L.V., Chmil V.S. and Shevchenko V.L., Experimental Study of Continuous Process of Manufacture of Lithium 12-Oxystearate Greases, in: Plastic Greases, Abstr. of Reports at 3rd All-USSR Sci. and Techn. Conf. (Berdyansk, September, 1979), Kaukova Dumka, Kiev, 1979, pp. 106-108.
24. Ishchuk Yu. L., Kuzmichev S.P., Krasnokutskaya M.E. et al., The Present State and Future Prospects for Advances in Production and Application of Anhydrous and Complex Calcium Greases, TsNIITEneftekhim, Moscow, 1980.
25. Grease Manufacturing: 1982 Refining Process Handbook, Bydroc. Proc., 1982, Vol. 61, No. 9, p. 212.
26. Ishchuk Yu. L., Fuks I.G. and Froishteter G.B., Production of Plastic Lubricating Greases, in: Diagram Manual of Production Layouts of Petroleum and Gas Processing, Ed. by Bondarenko B.N., Kliimiya, Moscow, 1983, Ch. 2, pp. 97-104.
27. Zaborskaya Z.M., Pavlov A.A., Ishchuk Yu. L. and Lendyel I.V., Specifics of Determination of Production Capacities of Grease Manufacturing Plants and Shops, Neftepererabotca i Nefteknimiya, Kiev, 1982, Issue 23, pp. 38-41.

Selection of Equipment for Individual Stages of Lubricating Grease Manufacturing Process

In the selection of equipment for manufacturing process stages, the rheologic and thermophysical characteristics of the intermediate products forming at individual stages of grease manufacture are taken into consideration first because they acquire specific properties [1-3]. As an example. Table 7.1 presents the properties of intermediate products in the manufacture of lithium 12-hydroxystearate greases, such as Litol-24. Similar characteristics for other types of greases and their intermediate products are described in [1]. Such properties as the thermal conductivity, heat capacity and density vary in the grease manufacturing process by not over 20% (Table 7.1).

Table 7.1. Change of properties of petroleum oil-lithium 12-oxystearate system in heating and thermal treatment

Temperature, °C	Viscosity		Density, kg/m³	Thermal conductivity, W/(m•K)	Heat capacity C_p, kJ/(kg•K)
	10 s^{-1}	100 s^{-1}			
Saponification in 2/3 of oil, water evaporation, and heating					
80	0.62	0.072	861	0.150	2.427
90	0.71	0.085	853	0.149	2.477
100	1.24	0.131	868	0.158	3.150
150	32.30	1.000	814	0.129	2.612
Charging of 1/3 of oil, heating, and thermal treatment					
150	2.42	0.48	821	0.133	2.701
180	7.40	1.11	790	0.122	2.800
210	5.02	0.68	762	0.111	2.980

The main parameters governing the specific requirements for the equipment used in the grease production process are rheological characteristics of greases and intermediate products, most important being their viscosity. LioSt-based and other soap greases, such as Uniol-type calcium complex ones, exhibit a considerable (by a factor of about 50-80) increase of the viscosity in the manufacturing process and its dependence on the strain rate [1]. These specific changes of rheological characteristics of greases in the course of their manufacture should be taken into account in calculations and selection of stirrers, pumps, heat exchangers and other units. Thus, in the selection of the stirrer for a reactor (cooking kettle) where the stages of mixing of the components, saponification (neutralization), dewatering, melt preparation and holding at the specified temperature and pre-cooling are combined, peculiar requirements are placed on its design. The requirements are best met by variable-speed scraper-and-paddle stirrers, which, allow different stirring conditions to be effected at different stages [2-8]. High efficiency of such stirrers coupled with the possibility of a more flexible control of the stirring rate promote shortening of the process time, saving of power, improvement of the quality of greases and reproducibility of their properties in manufacture of individual batches.

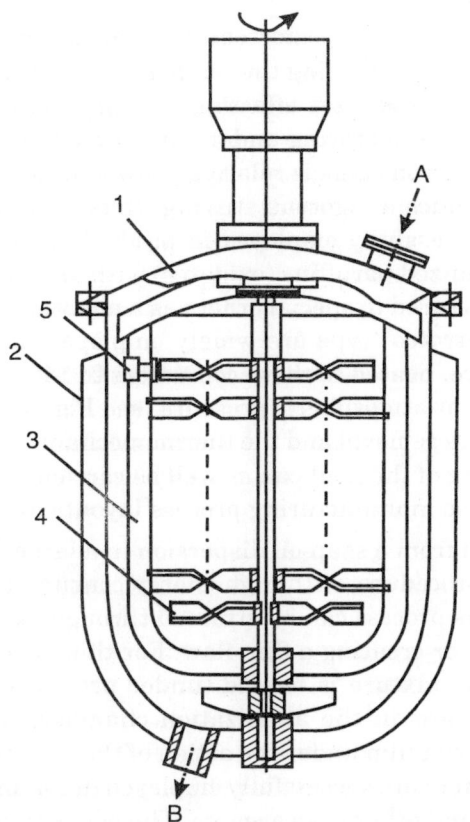

Fig. 7.1. Reactor with a scraper-and-paddle stirrer (V =10 m^3):
1—housing; 2—jacket; 3—frame; 4—paddle for mixing;
5—scraper with fixture;
A—charging of raw materials;
B—discharge of a product.

Figure 7.1 shows a reactor of 10 m^3 capacity with a scraper-and-paddle stirrer; such units are also manufactured with capacities of 5 and 16 m^3. The heat-exchange surface of the reactor ($V = 10$ m^3) is 20 m^2. Its stirrer consists of a central five-row radial-paddle mechanism (rotation speed $n = 1$ and $0.5\ s^{-1}$) and a peripheral mechanism, a scraper frame with four rows of scrapers, rotating in the opposite direction ($n = 0.33$ and $0.17\ s^{-1}$). The scrapers are hydrodynamically pressed against the inside surface of the reactor wall. For a better fit to the surface the scrapers are composed of separate section in the height. The cutting part of a scraper is made from a fluoroplastic in order to avoid galling. The stirrers are driven from a 40-kW two-speed gearmotor through a special transfer case with two coaxial output shafts providing for opposite directions of rotation of the central and peripheral stirrers as well as the possibility of a separate operation of either of them. Such a design of stirrers significantly enhances the heat and mass transfer and reduces the temperature difference of the reactor content from the reactor wall to the central shaft. The maximum temperature difference in both heating and cooling cycle is not over 12°C, whereas in units with an ordinary frame stirring it reaches more than 40°C [6,7].

The cycle time can be cut down by shortening the duration of the stage of saponification of natural fats (triglycerides) by increasing the temperature in the apparatus. The presence of water in the reaction mass necessitates effecting this process under pressure. It is accomplished in contactors-autoclaves operating under a pressure upto 1 MPa [2-13]. The viscosity of the system at the saponification stage is relatively low and therefore highspeed stirrers can be used in contactors to produce a vigorous stirring; thus autoclaves fitted with high-speed turbine-type stirrers are successfully employed in production of synthetic fatty acids at the stage of saponification of oxidized paraffins (oxidates) with an aqueous solution of NaOH [14]. Such apparatus are manufactured to special orders as a nonstandard equipment. Similar contactors-autoclaves of the "Stratco" type are widely employed abroad in the manufacture of soap greases [8]. The mixture, heated in these contactors to 150-170°C, is after saponification sent into reactors operating at atmospheric pressure (see Fig. 7.1). The additional heating in the reactor results in moisture removal and the thermomechanical dispersion of the soap thickener formed. A partial cooling of the melt can as well be carried out when required. Some data on the reactors used in grease manufacturing process layouts are presented in Table 7.2.

The moisture removal from a soap-oil dispersion in a large volume of a reactor is a time consuming and ineffective procedure, with high energy consumption. Shortening of the duration and improvement of this process can be attained through extending the evaporation surface by atomizing the liquid or creating a film flow. For this purpose an evaporator has been developed (2, 15) where the mixture is heated (under pressure) to 150-170°C and a major amount of moisture evaporates in the atomization chamber, which is followed by a deep dewatering in the flowing-down film under the action of the heat input through the wall of the unit (Fig. 7.2). Such evaporators are successfully employed in commercial production of lithium, complex, hydrated calcium and other soap greases. The capacity of the evaporator in terms of the moisture evaporated from the reaction mixture in the manufacture of greases is about 170 kg/h. The maximum permissible temperature of the wall of its shell is 200°C, and of the heat carrier—280°C. The evaporator can operate under a vacuum down to 90%. The maximum permissible pressure in the evaporator jacket is 0.6 MPa, and in its coil, 1.6 MPa. A special design of the distributing device of the evaporator makes it possible to control over a broad range the proportion of the moisture removed from the mixture in the flowing down film in accordance with the viscous properties of the liquid being dewatered.

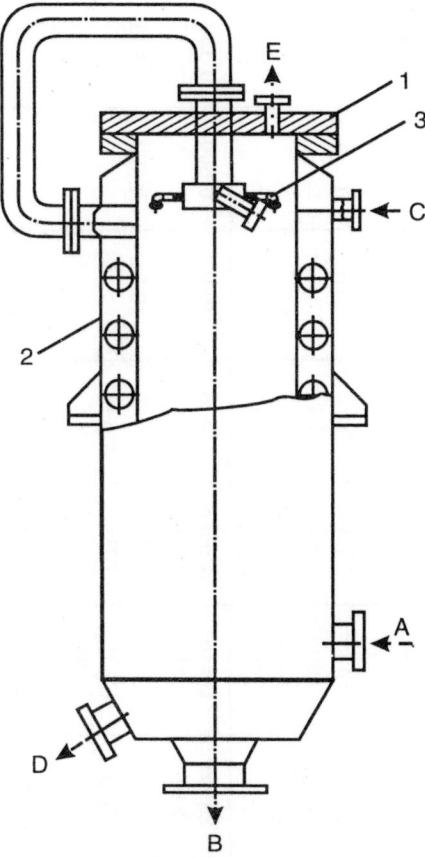

Fig. 7.2. Evaporator AV-600:
1—housing; 2—jacket; 3—spraying device;
A—charging of a product; B—discharge of a product;
C—heatcarrier supply; D—heatcarrier discharge;
E—discharge (suction) of vapours.

The heating (thermal treatment) or cooling of the high-viscosity soap-oil concentrate in reactors (see Fig. 7.1) is inefficient as well. It is advantageous to use for this purpose heat exchangers of a special design [2-5, 8, 10, 11]. Heat exchangers of a scraper type, where a continuous cleaning of the surface by scrapers eliminates the adverse effect of an increased adhesion of the product and its viscosity is lowered due to application of high shearing stresses, proved themselves to be advantageous in grease manufacture, Fig. 7.3 shows a two-shell scraper-type heat-exchanger surface of 3.5 m² [2, 3]. Each shell accommodates a rotating shaft with four "floating" scrapers which under the action of the centrifugal force and of the pressure of the liquid being stirred press tightly against the inside chrome-plated surface of the cylinder; the latter is of 0.35 m in diameter and 1.67 m in length and has a useful (free) volume of 0.05 m³. The individual drive is through a speed reducer; the rotation speed, in the first (counted in the direction of the product flow in cooling) shell (cylinder) is $n = 4$ rpm and in the second, 3 rpm the rating of the electric motor for the first shell is 17, and for the second shell, 22 kW. When heated the product flows through the shells in the reverse order. Pre-cooled water, purified of

Table 7.2. Characteristics of reactors used in manufacture of greases

Reactor, function (manufacturer)	Type, model	Volume, m^3	Temperature, °C		Pressure, MPa		Motor rating, kW	Rotation speed, s^{-1}	
			in apparatus	in jacket	in apparatus	in jacket		paddles	scrapers
With turbine stirrer; mixing of components and neutralization (saponification) of acids (fats) without water evaporation (Minkhimmash USSR)	Enclosed type	5 10 16	Up to 250 Up to 250 Up to 250	Up to 250 Up to 250 Up to 250	Up to 0.6 Up to 0.6 Up to 0.6	Up to 0.4 Up to 0.4 Up to 0.4	10.0 22.0 22.0	2.160 2.200 2.200	— — —
With frame-type stirrer; mixing of components and neutralization (saponification) of acids (fats), water evaporation, thermal treatment, partial cooling (Minkhimmash USSR)	Enclosed type	5 10 16	Up to 250 Up to 250 Up to 250	Up to 250 Up to 250 Up to 250	Up to 0.6 Up to 0.6 Up to 0.6	Up to 0.4 Up to 0.4 Up to 0.4	5.5 7.5 10.0	0.750 0.535 0.535	— — —
With scraper-and-paddle stirrer; mixing of components and neutralization (saponification) of acids (fats) water evaporation, thermal treatment, partial cooling (Mink-himmash USSR)	RS 5 RS 10	5 10	Up to 210 Up to 210	Up to 280 Up to 280	Atmospheric Atmospheric	Up to 0.4 Up to 0.4	30.0 40.0	0.800 0.800	0.330 0.330
Reactor-contactor (autoclave); mixing of components, saponification of fats under pressure ("Simmering Gras Pouker", Austria)	"Stratco"	7.8	260	300	1.0	—	—	—	—
With scraper-and-paddle stirrer; mixing of components and neutralization (saponification) of acids (fats) water evaporation, thermal treatment, partial cooling ("Simmering Gras Pouker", Austria)	—	10	180	Up to 230	Atmospheric	1.0	65.0	0.500 0.250	0.330 0.160

impurities and salts, is fed into the jacket of the unit (in each shell separately) for cooling, and oil or another heat carrier, for heating. The capacity of the unit in cooling of greases in the range of temperatures from 200 to 60°C is with a flow rate of 2000-2500 kg/h the cooling water with an initial temperature of 5°C and can be increased up to 20 m³/h. The capacity of the unit in heating is 20-25% higher than in cooling provided the grease-heat carrier temperature difference remains the same. The permissible working pressure in the unit is up to 2.5 MPa, and in its jacket, up to 0.6 MPa. The heat transfer coefficient in these units in cooling of lithium greases amounts to 600-650 W/(m². K), which is about 20 times as high as in ordinary tubular heat exchangers. This unit is, however, not efficient for heating of greases and is recommended only for their cooling.

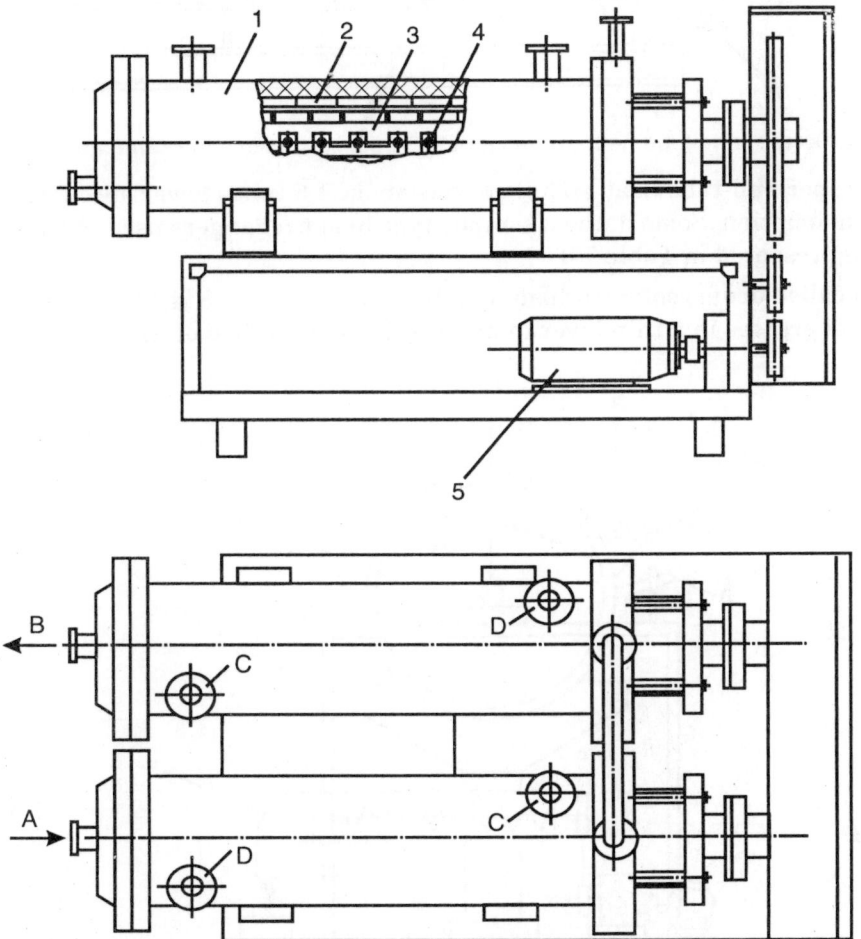

Fig. 7.3. Scraper-type heat exchanger GS-350:
1—housing; 2—jacket; 3—shaft; 4—scrapers; 5—shaft drive;
A—charging of a product; B—discharge of a product;
C—heatcarrier supply; D—heatcarrier discharge.

The scraper-and-tube heat exchanger [2, 3, 16] shown in Fig. 7.4 is more attractive since it is efficient for heating and cooling of greases in continuous processing plants. Its heat-exchange surface is of 3.75 m², including that of the cylinder of 1.25 m²; capacity, from 2 to 2.5 t/h; working pressure in the jacket, up to 0.6 MPa, and in the shell, up to 2.5 MPa; electric motor rating, 17 kW; and scraper shaft rotation speed, 4/sec. The heat carrier temperature (in heating) is up to 250°C and the coolant temperature (in cooling), 5-15°C; the product temperature at the outlet is up to 220°C in heating and down to 40-45°C in cooling.

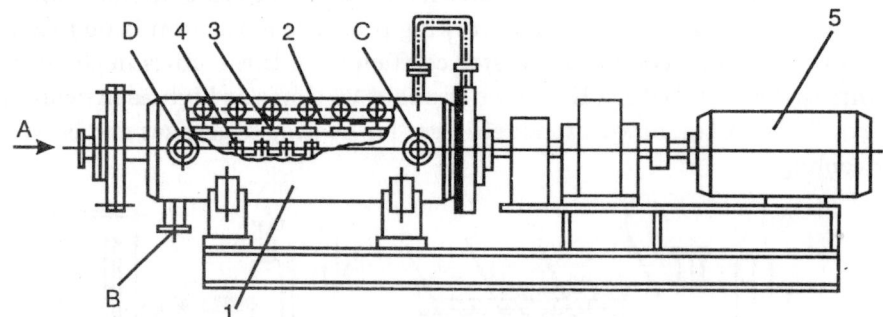

Fig. 7.4. Scraper-and-tube heat exchanger GSG-250; inscriptions the same as for Fig. 7.3.

The scraper-and-tube heat exchanger has about 1.5 times lower specific metal content and power consumption. Some data on scraper-type heat exchangers used in the manufacture of greases are presented in Table 7.3.

The so-called drum coolers, schematically shown in Fig. 7.5 [4,11-13], are employed for rapid cooling of greases in a thin layer in an intermittent method of their manufacture, most

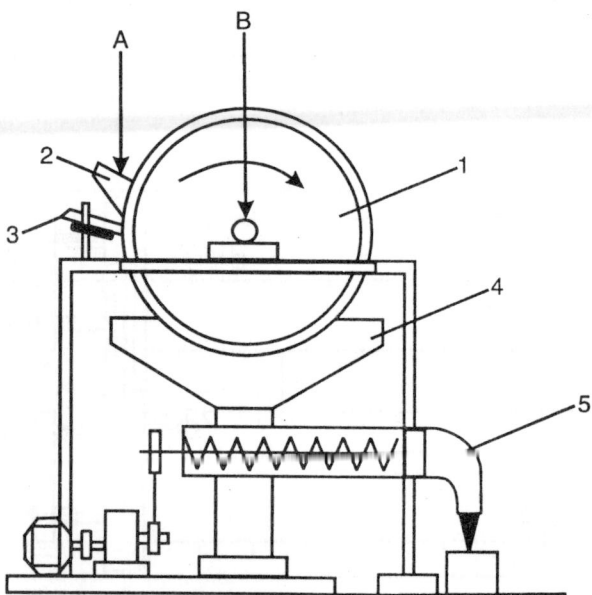

Fig. 7.5. Drum-type cooler:
1—cooling drum; 2—chute for grease feeding;
3—blade for grease shaving off;
4—receiving bin; 5—screw feeder for grease filling;
A—charging of a product; B—supply of coolant (water).

Table 7.3. Characteristics of scraper-type heat exchangers used in manufacture of greases

Heat exchanger, function (manufacture)	Model	Heat exchange surface, m^2	Capacity, t/h	Drive rating, kW	Number of shells	Rotation speed of scrapers in shells/sec		
						1	2	3
Scraper-type; cooling from 200-210 °C to 45-50°C (Minkhimmash USSR)	TS-250 TS-250	2.50 3.50	2.0-2.5 3.0-3.5	total, 30 total, 39 (17+22)	2 2	4 4	3 3	3 -
Scraper- and -tube-type; heating and cooling over broad temperature range (up to 250 °C) (Minkhimmash USSR)	TS-250	3.75 (including 1.25 of scraper device)	2.0-2.5	17	1	4	-	-
Scraper-type; heating and cooling in temperature range of 50-150 °C ("Johnson, Great Britain)	262CX	2.25	Up to 5.0	28	2	3	4	-
The same, in temperature range of 50-200°C	373 CCX	3.50	Up to 5.0	28	3	3	4	4

often for cooling Constalin-type sodium greases or hydrocarbon greases. The operating principle of such a cooler is that the grease melt from the cooking kettles is fed into chute 2 that tightly bears against the surface of rotating cylinder-drum 1, having a hollow shaft fed with the coolant, water. The grease melt from the chute is uniformly distributed over the surface of the drum (d = 1.5-2 m; l = 2-2.5 m), cooled in a time equal to one revolution of the drum, shaved off by blade 5 mounted under the chute, and falls down into bin 4, from which it is fed by screw feeder 5 for filling into containers. The required grease cooling conditions are created by varying the drum rotation speed, thickness of the grease film being cooled, feed rate and temperature of the coolant.

The homogenisation (mechanical working) of greases under laboratory and commercial production conditions is effected by various methods [2-5, 8-13, 17, 18]. They are often pressed through a set of screens or screen filters with various mesh sizes of the wire cloths used; sometimes are pumped through a system of rotating perforated disks or stirred in stirrers from a penetrometer; efficient mixers as well as three-roller mills are employed or a repeated circulation of some greases is effected with the aid of rotary gear or screw pumps; devices based on the ultrasonic effect and other methods are used. Many of them are of a very low efficiency.

Under laboratory conditions a good result is attained at homogenisation of greases in three-roller mills (paint grinders) with an adjustable roll gap. Well operating colloid mill-type homogenizers manufactured by "Koruma" (FRG) [21] as well as "Monton Gaulin" valve-type homogenizers (USA) [22] or domestic valve-type homogenizers manufactured by the Odessa food-industry Equipment Works, have been recently used in the commercial production at domestic factories. Colloid mills are employed mainly in intermittent grease manufacture, while valve-type homogenizers are used in continuous processes [19].

The homogenisation of greases in colloid mills is effected in an adjustable gap between stator and rotor disks under the action of very high shear stresses. Industrial-type machines are manufactured with a capacity from 0.5 to 6 t/h, which varies insignificantly within permissible limits depending on the diameter of disks, amount of gap between them, grease composition and viscosity, and rate of grease feed into the homogenisation zone. The diameter of disks is usually within 60 and 350 mm and the gap between them is adjustable from 0.06 to 2 mm. The "Koruma" homogenizers can be fitted with devices (plants) for the grease filtration and deaeration, whose capacity corresponds to that of the homogenizer. An USSR-make of colloid mill homogenizer is shown in Fig. 7.6.

Valve-type homogenizers are high-pressure plunger pumps (industrial, mainly three, and laboratory, single-plunger ones) with homogenizing heads-valves [20,22]. Grease under high pressure is fed into the homogenizing head and flows at a high velocity through a narrow adjustable annular gap between lapped surfaces of the homogenizing valve and its seat. Valve-type homogenizers with a throughput capacity of 2-3.5 t/h at the maximum pressure up to 50 MPa are usually employed in the commercial production of greases. The annular gap varies within 0.05 and 0.8 mm depending on the pressure. Some data on valve-type homogenizers and colloid mills, which can be used in grease processing, are presented in Table 7.4.

Table 7.4. Characteristics of homogenizers used in manufacture of greases

Homogenizer, function (manufacturer)	Model, Type	Capacity, kg/h	Rotation speed of disks, s^{-1}	Maximum pressure, MPa	Maximum temperature, °C	Mass, kg	Overall dimensions, mm		
							length	width (diameter)	height
Valve-type; mechanical treatment (homogenization) of semiproducts and greases (Odessa food industry Equipment Works)	OGM-10	Up to 10,000	—	25	85	—	—	—	—
	Al-OGM	Up to 5,000	—	20	85	1710	148.0	111	164
	K5-02A1,2	Up to 1,200	—	20	85	850	86.5	93	140
As above ("SOAVIB. FIGI", Italy)	B-63	Variable up to 10,000	—	50	80	—	240	200	220
	DA-61	6,000	—	50	80	—	150	160	200
	D-61	4,000	—	50	80	—	150	100	130
	CA-60	3,000	—	50	140	—	130	90	110
	A-58T	2,000	—	50	80	—	200	115	162
	A-58	1,000	—	50	80	—	100	76	140
	X-68	500	—	50	80	—	70	60	85
	Z-57	100	—	50	80	—	60	45	65
As above ("Alfa-Laval", FRG and Sweden)	H-35	10,000-25,000	—	20	80	—	150	190	185
	H-30	4,000-12,000	—	35	80	—	148	184	180
	H-20	500-4,000	—	35	80	—	95	110	137
	H-10	50-500	—	63	80	—	85	110	122
As above ("Gaulin Corporation", USA)	MC18-2TR	13,000	—	14	80	2100	150	150	124
	MC-2TR	5,600	—	14	80	1300	114	152	104
	31M-3TA	-	—	25	100	—	81.8	43.2	111.8
	15M-3TA	-	—	70	100	—	81.8	43.2	111.8

As above ("A.P.V. Company Ltd. Monton Gaulin", Great Britain)	K24-3	4,500-9,000	—	21 and 35	90	4140	193	128.3	113.4
							152.4	118.8	111.1
	K12-3	2,500-4,500	—	21 and 35	90	2370	127	99.1	100.3
	K6-3	900-2,500	—				99.1	83.7	97.2
	K3-3	270-900	—	21 and 35	90				
	15M-8BA	about 60		21 and 35 56	90 100	1550 682			
Colloidal mill; mechanical treatment (homogenization of semiproducts of greases (Minkhimmash USSR)	250	Up to 4,400	50	—	80	1160	239	75	222
As above ("Koruma", FRG)	5	Up to 6,000	50	—	80	—	—	—	—
	4	Up to 4,000	50	—	80	850	258	70	215
			50	—	80	690	230	65	200
	3	Up to 2,500	50	—	80	—	—	—	—
	2	Up to 650	50	—	80	—	—	—	—
	Laboratory	4-50	50						
As above ("Gaulin Corporation", USA)	8B	700-7,000	166.6	—	100	—	140	46	40
	4	300-2,800	166.6	—	100	—	87	36	31
	2	90-280	166.6	—	100	—	65	31	28

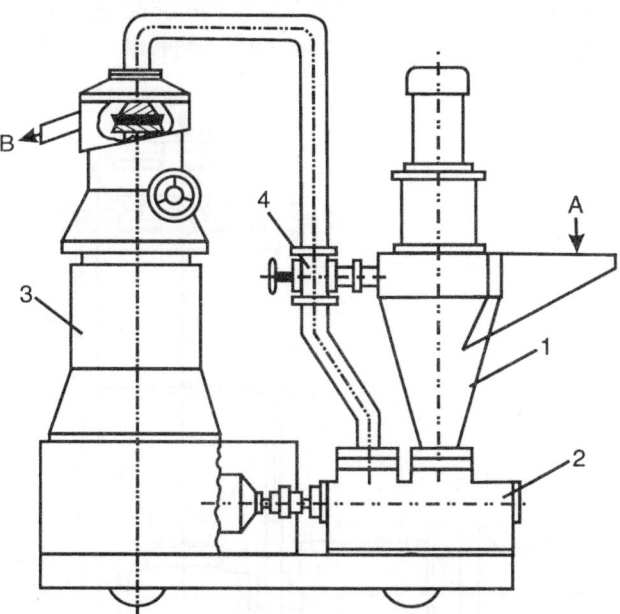

Fig. 7.6. Homogenizer "colloid mill" G-200:
1—bin; 2—pump; 3—mill (milling disks with ajustable gap up to 0,05 mm); 4—three-way valve;
A—charging of a product; B— product discharge.

Greases are sometimes homogenized at domestic factories with the use of "Leman"-type homogenizers, which are three roller mills with a roller diameter of 400 mm [4]. The grease homogenisation occurs in an adjustable gap between the rollers. These homogenizers are employed only in an intermittent production method and have a very high specific power consumption. The lowest power consumption is offered by valve type homogenizers.

Deaeration of greases can be carried out continuously in a unit whose design is shown in Fig. 7.7 [23, 24]. The throughput capacity of the deaerator varies within 1-3 t/h depending on the capacity of the grease feeding and pumping out pumps, vacuum in the deaerator, size of slots between plates, through which the grease is pressed, as well as on the grease deaeration degree and other grease characteristics, such as viscosity. The grease enters the deaerator through pipe connection A, passes through adjustable slot gaps (0.1-0.5 mm) between plates, and gets into the vacuum chamber ($V = 0.16$ m^3) as thin ribbons. Vacuum in the unit is produced through pipe connection C, and the deaerated grease is continuously withdrawn from the bottom of the unit by a rotary gear pump, such as RZ-30I, driven through a variable-speed reducer (rotation speed variation range, 1-100 s^{-1}) from a 4-kW electric motor. The rotation speed is adjusted so as to match the grease inlet and outlet flow rates. A practically complete air removal is attained in such a deaerator at a vacuum of 90-95%.

In deaerators delivered in the set of the "Koruma" homogenizers [21, 25] the grease is pumped by a high-speed screw pump into the deaerating chamber through perforated plates, one of which is movable, while the other is stationary. When perforations in the stationary plate are closed, vacuum is created, and air from the grease which, having passed through perforations in the plates, is in the form of threads is removed into the discharge piping. Data on deaerators used in the manufacture of greases are presented in Table 7.5.

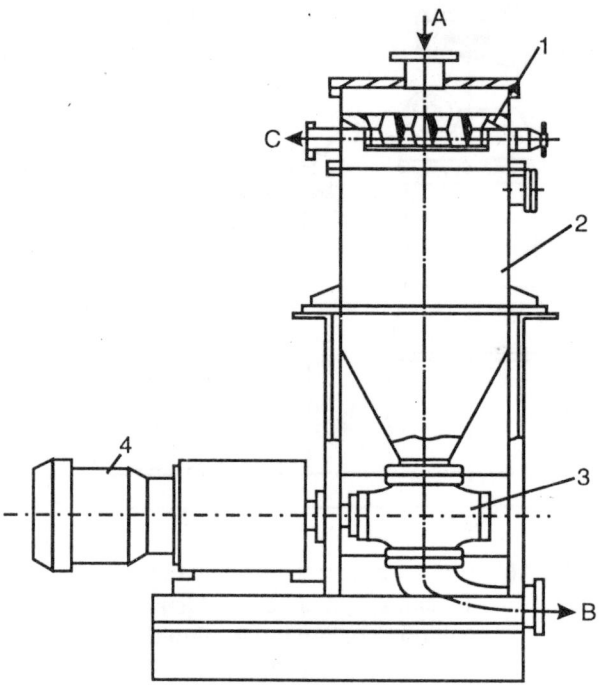

Fig. 7.7. Vacuum deaerator:
1—slot device; 2—housing; 3—pump; 4—pump drive;
A—charging of a product; B—product discharge;
C—air discharge (suction).

Modern grease manufacturing processes, especially continuous ones, employ highly sensitive proportioners, high-speed mixers, homogenizers with small gaps between working elements, devices for monitoring and control of the processing conditions as well as for an in-process inspection of greases, and therefore filtration of starting components, intermediate products, and finished greases becomes exceedingly important. Various oils and other liquid components are filtered in separating and filtering centrifuges, filter presses, drum filters operating under pressure and vacuum as well as in other devices [4, 7, 9, 11, 26]. Filtration of finished greases and intermediate products of their manufacture is more complex. Greases are often filtered in screen filters of various designs, but such filters get clogged, require periodic cleanings, and are therefore unsuitable for filtration of greases in continuous processes, It is more convenient to employ for this purpose either drum filters with several filter elements or cartridge-slot mechanized filters of the "FDZh" type [26].

"Purolejector"-type filters [10,11] with changeable filter elements are the most effective among mechanized filters for the filtration of greases. They comprise of hollow perforated cylinders wound around with a wire so that the gap between wire turns is from 0.025 to 0.5 mm. Grease is filtered by pressing it through the gaps. The filter elements are changed in

Table 7.5. Characteristics of deaerators used in manufacture of greases

Deaerator, function (manufacturer)	Model	Capacity, kg/h	Vacuum chamber volume, l	Vacuum depth, %	Overall dimensions, mm		
					length	width (diameter)	height
Vacuum-type; air removal from grease (Minkhimmash USSR)	D-400	300	160	96	1900	1155	2240
Vacuum-type; air removal from grease ("Simmering Graz", Austria)	—	—	180	90	1600	1300	2670
Vacuum-type; air removal from grease ("Koruma", FRG)	EVR/2	120	—	90	1450	550	1400
	EVR/3	380	—	90	1800	620	1650
	EVR/4	1500	—	90	2200	700	1800
	EVR/5	3000	—	90	2450	700	2150
	EVR/6	5000	—	90	2600	700	2150

Note. The deaeration degree reaches 100%.

Table 7.6. Characteristics of filters used in manufacture of greases

Filter; function (manufacturer)	Model	Filtration surface, m^2	Capacity, t/h	Working pressure, MPa	Gap between plates, mm
Cartridge-slot mechanized filter; Filtration of components, semiproducts, and greases (Minkhimmash USSR)	FDZh-80 FDZh-100 FDZh-150	0.033 0.066 0.132	Up to 1.8 Up to 3.6 Up to 7.2	0.10 0.12 0.15	0.18 0.18 0.18
Mechanized filter with changeable filter elements (pyrolejtor); filtration of components, semiproducts and greases (FRG)	G-141	—	0.7-4.0	1.00	0.025-0.50
Filtering plant; filtration of components, semiproducts, and greases ("Koruma", FRG)	2 2A 3 4 5	— — — — 	0.65 1.50 2.50 4.00 Up to 6.00	Up to 0.5 Up to 0.5 Up to 0.5 Up to 0.5 Up to 0.5	0.2 0.2 0.2 0.2 0.2

accordance with the required filtration quality. The filters are cleaned by rotating the filter element around a fixed blade.

Filtering plants manufactured, by "Koruma" or "Frima" proved themselves effective in intermittent and semicontinuous processes [10,11,21,25].

Table 7.7. Characteristics of mixers used in manufactureof greases

Mixer, function (manufacturer)	Model	Characteristics
Rotary hydrodynamic of flow-through type; mixing of components (Minkhimmash USSR)	GART-Pr	Capacity of water, up to 80 m^3/h; working pressure, up to 0.8 Pa; product temperature, up to 90°C; rotor-radiator diameter, 280 mm; rated rotor-radiator rotation speed, 25 s^{-1}; electric motor rating, 20-55 kW
	GART-PrM	Capacity of water, up to 2.5 m^3/h; working pressure, up to 1.0 mPa; product temperature, up to 90°C; rotor-radiator diameter, 140 mm; rated rotor-radiator rotation speed, 50/sec; electric motor rating, 4 kW
With vortex layer; mixing of components (Minkhimmash USSR)	VA-100	Capacity of water: up to 15 m^3/h [28], up to 20 m^3/h [29]; working zone diameter, 70 mm; voltage, 380-220 V; frequency, 50 Hz; power consumed: 1.6 kW [28, 9], 3 kW [20]
	AVS-100	Capacity of water up to 15 m^3/h [28]; working zone diameter, 70 mm; voltage, 380-220 V; frequency, 50 Hz; power consumed: 1.6 kW
	AVS-150 or AVS-150V*	Capacity of water up to 40 m^3/h [27], up to 30 m^3/h [30]; by greases, 7 m3/h [30]; working zone diameter: 128 mm [28], 150 mm [30]; voltage, 380-220 V; frequency, 50 Hz; power consumed: 56 kW [28], total 63 kW and active 9.5 kW [30], 3.5 kW [10]
Flowthrough type; mixing of components and soap concentrate with remaining part of oil ("Sil-Verson", Great Britain)	Silverton	Capacity of water up to 10 m^3/h; gap between rotor and stator, 0.3 mm; three-speed drive; rotation speeds, 47,2, 23,9, and 15.8/sec; power consumed, 4,1, 2.05, and 1.35 kW respectively

Some data on filters suitable for filtration of components, greases and intermediate products in their manufacture are presented in Table 7.6.

An important procedure in the modern production of greases is the mixing of starting components and semiproducts at individual manufacturing stages. Specific features of the mixing in reactors (kettles) in the grease manufacture by the intermittent or the semicontinuous method have been discussed above. It is very important, however, to attain an effective mixing of components in a continuous grease manufacturing process. Used for this purpose are mixers, some data on which are presented in Table 7.7.

The GART-Pr rotary hydrodynamic flow mixers [27] as well as the VA and AVS mixers with a vortex layer of ferromagnetic particles [28] provide a highly effective mixing of oils, fats, and hydroxides in the manufacture of soap greases, the mixing being accompanied by a partial saponification of fats and, when free fatty acids HSt, 12-HoSt, HOI are used, by their neutralization [10]. The GART-Pr and AVS (VA) mixers are schematically shown in Figs 7.8 and 7.9 respectively.

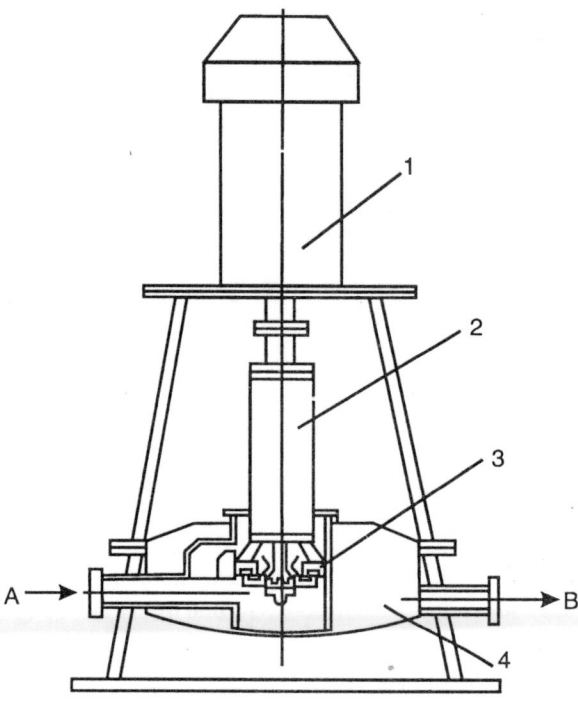

Fig. 7.8. Hydrodynamic flow mixer GART-Pr:
1—electric motor; 2—rotor head; 3—rotor emitter; 4—working chamber.
A. Inlet
B. Outlet

Various sources [28-50] present substantially different technical data for one and the same mixers with a vortex layer of ferromagnetic particles, such as for VA-100 and ASV-150 (see Table 7.7), the discrepancies being particularly high for electric power parameters, namely the power consumed. Because of this, the data on the specific power consumption, specific metal content and other technical and economic characteristics should be dealt with rather critically.

Apart from the units listed, in Table 7.7 other mixers can also be used to intensify the mixing processes, such as a flow mixer with a throughput capacity of upto 7 t/h with a variable three-stage rotation speed. (16.7; 25.0; and 50 s^{-1}) of the rotor (2); centrifugal turbine-type mixers (8); ultrasonic mixers (31); mixers using the energy of laser devices (32), etc.

In the manufacture of greases, especially by semicontinuous and continuous processes, it is important to select correctly the pumps. The transfer of such components as dispersion media (petroleum and ester oils, polysiloxanes, and other liquids), saponifiable components (vegetable oils, salomases, fatty acids) is effected by ordinary centrifugal, gear or screw pumps, manufactured by the domestic industry. Special gear, screw, and other pumps are often used for the transfer of greases and especially of intermediate products in their manufacture. Some data on pumps and metering pump units, used in the manufacture of greases, are presented in Table 7.8 [33].

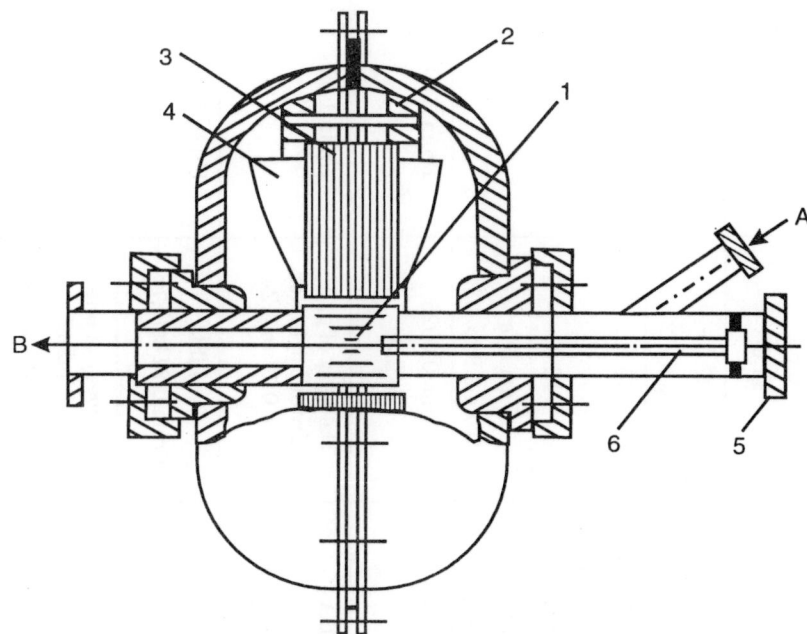

Fig. 7.9. Mixer with vortex layer of ferromagnetic particles:
1—reaction chamber; 2—housing; 3—cavity-type tip;
4—winding; 5—cover; 6—holder;
A—charging of a product; B—product discharge.

Accumulating tanks (Fig. 7.10) are installed to ensure continuity of the main grease manufacturing process before the filling of greases into delivery containers. The number of such accumulating tanks depends on their volume and throughput capacity; it is advisable that three such tanks be provided: one, under loading; the second, in averaging the quality of a finished grease and under laboratory check; and the third, in unloading, *i.e.*, packaging of the finished grease. The accumulating tanks have a conical shape (14) which facilitates the discharge of grease from them. They are often fitted with a water jacket for a moderate heating of the grease in order to reduce its sticking to the tank wall and to reduce its viscosity with the aim of facilitating the operation of the pump installed at the tank bottom to circulate the grease through the accumulating tank—pump—accumulating tank circuit and to transfer it to units for filling the delivery containers. Screw pumps are commonly used to circulate and to pump out the grease.

Table 7.8 Characteristics of pumps and metering units in manufacture of greases

Pump, function (manufacturer)	Model	Delivery, m^3/h	Pressure, MPa	Rotation speed, s^{-1}
Three-screw pump; transfer of semiproducts and greases up to 80°C (with appropriate modification of seals, up to 200°C) (Minkhimmash USSR)	3V4/25	3.2 6.6 11.0	0.50 2.50 2.50	24.1 48.2 24.1
Three-screw pump; transfer of semiproducts and greases up to 200°C ("Allweler")	3V16/25	22.0 Up to 33	2.50 Up to 1.00	48.1
Gear pump; transfer of semiproducts and greases (Minkhimmash USSR)	ShG-20/25A up to 190°C RZ-30 up to 80°C	16.5 16.0	Up to 1.00 Up to 0.60	24.1 24.1
Gear pump unit; transfer of semiproducts and greases up to 190°C (Minkhimmash USSR)	ShG20/2.5 Rp-1	Variable from 0.8 to 4.3		1.4-8.3
Metering plunger pump; metered transfer of stock components, semiproducts, and greases up to 200°C (Minkhimmash USSR)	ND	From 0.0025 to 2.5	From 1.00	Up to 40.0
Metering pump unit; 2-, 4-, and 6-plunger pumps; metered transfer of stock components, semiproducts, and greases (Minkhimmash USSR)	DA	Delivery of one hydraulic cylinder, from 0.17 to 4.24	Up to 1.00	
Eccentric cam pump; transfer of semiproducts and greases up to 150°C ("Duglus Rounson", Great Britain)	EH	Variable From 0.21 to 7.76	Up to 1.75	
Polyplunger metering pump; metered transfer of stock components and greases up to 300°C ("Lewa", FRG)	Multiplunger metering unit, metered transfer of stock components and greases up to 300°C ("Yewa", FRG)	Delivery of one cylinder, from 0.007 to 2.85	From 0.5	Up to 300

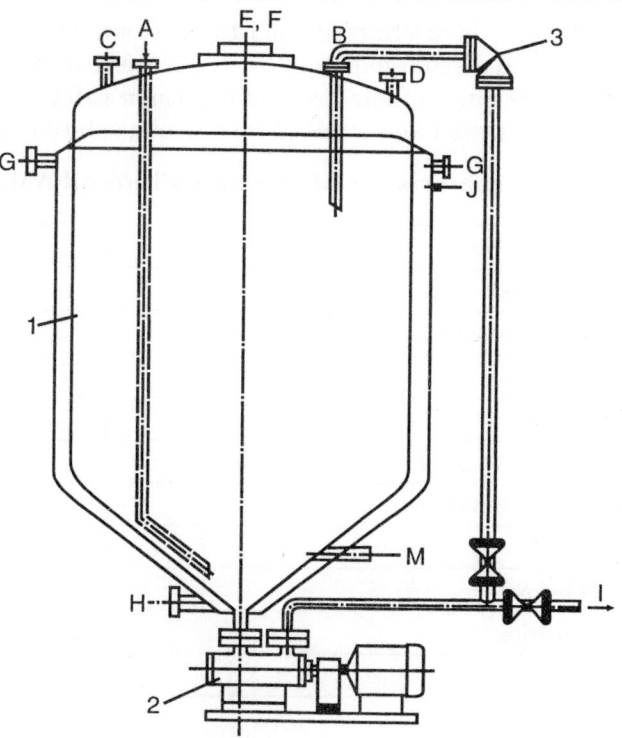

Fig. 7.10. Accumulating tank:
1—housing; 2—pump; 3—homogenizing valve;
4—A—charging of a product; B—recirculation of a product;
C—compressed air supply; D—air discharge; E—manhole;
F—inspection window; G—hot water supply;
H—water discharge; I—finished product discharge (grease);
J—outlet of high level indicator; M—outlet of low level indicator.

The specified temperature conditions in the grease manufacturing process are provided with the use of various coolants and heat carriers, fed into special jackets or coils of individual apparatus, pipings, pumps, and other equipment units [8, 10-13, 35, 36].

Saturated steam is the most safe and efficient heat carrier, but the possibilities of its use are restricted by the fact that jackets of main processing units are mostly rated for a pressure not over 0.6 MPa and hence their contents cannot be heated above 140-145°C. It follows that saturated steam of a pressure upto 0.6 MPa can be used as the heat carrier in relatively low-temperature grease manufacturing processes, which include, *e.g.*, manufacture of hydrated and anhydrous Ca-greases, ordinary Al-greases as well as greases based on hydrocarbon, silica gel, bentonite and other inorganic thickeners.

Steam to grease manufacturing plants is usually supplied centrally from boiler plants of factories or from separate heat-and-power plants. Waste (exhaust) steam or condensate as well as hot water are used in some cases for heating individual equipment units.

The use of steam for heating of units intended for both heating and cooling their contents allows the use of water as the coolant. Steam is always fed into the jacket of an apparatus from the top and its condensate is discharged from the bottom, whereas water, on the contrary, is fed from the bottom and discharged from the top. The efficiency of steam and water as the heat carrier and coolant is due to their fire and explosion safety, nontoxicity, high heat transfer coefficient, high evaporation and condensation heats, relative cheapness, etc.

Table 7.9. Characteristics of Thermic Fluid oil AMT-300

Characteristics	Value
Density at 20°C, kg/m^3 (not below)	960
Kinematic viscosity at 100°C, mm^2/s (not over)	5.9
Refractive index n_D^{20} (not below)	1.540
Pour point, °C (not over)	-30
Flash point in closed crucible, °C (not below)	170
Flash point in open crucible, °C (not below)	194
Ignition point, °C (not below)	220
Self-ignition point, °C (not below)	285
Content of mechanical impurities	absence
Acid number, mg KOH/g (not over)	0.03
Fractional composition, °C:	
5% boils off (not below)	330
95% boils off (not over)	475
Range of ignition of vapor in air, g/m^3	140-400
Maximum permissible concentration in production room air, mg/m^3	300
Lower limit of explosive aerosol mixture, g/m^3	25
Maximum explosion pressure, MPa	0.74

When, however, a temperature rise up to 200°C and over is needed, such as in the manufacture of Na-, Li-, cCa-, cAl-, c/Li-, and other greases, the use of steam (even superheated one) is either ineffective or impossible, and other heating methods are resorted to. Steam is substituted with organic heat carriers, such as a diphenyl mixture consisting of diphenyl oxide (73.5%) and diphenyl (26.5%), petroleum oils, special thermally stable highly aromatized petroleum fractions, and other high-temperature organic heat carriers (HOH) [2-13,35,36]. Used in the domestic practice as a high-temperature petroleum-origin heat carrier is thermic fluid AMT-300 or AMT-300T 35 (Table 7.9), produced according to (75).

Organic heat carriers are heated to 250°C and above in various heaters and to TU 38-101537- heating units [8, 12, 13, 36-38]. When the diphenyl mixture is used, special diphenyl boiler plants are constructed, which have, however, not found wide use in the domestic grease manufacture practice. Petroleum oils and other HOH, when used as heat carriers, are heated in electric heaters or special radiant, convective, or radiant-convective furnaces, fired with a gaseous or a liquid fuel. Electric heaters and small-size tubular furnaces are serially produced

Table 7.10. Characteristics of electric heater

Electric heater type	Total power, kW	Thermal output, MJ/h	Total capacity of electric heater, m^3	Total number of heating elements in electric heater	Power of one heating unit, kW[1]	Number of heating units	Power of one heating unit, kW	Number of groups in one heating unit	Power of groups in heating unit, kW			
									1st group	2nd group	3rd group	4th group
Three-section	$\frac{604}{378}$	$\frac{1400}{1040}$	1.5	252	$\frac{2}{1.5}$	3	$\frac{168}{126}$	4	$\frac{84}{63}$	$\frac{42}{31.5}$	$\frac{18}{13.5}$	$\frac{12}{9}$
Two-section	$\frac{336}{252}$	$\frac{920}{700}$	1.0	168	$\frac{2}{1.5}$	2	$\frac{168}{126}$	4	$\frac{84}{63}$	$\frac{42}{31.5}$	$\frac{18}{13.5}$	$\frac{12}{9}$
	$\frac{128}{96}$	$\frac{360}{264}$	0.4	64	$\frac{2}{1.5}$	2	$\frac{64}{48}$	$\frac{3}{2}$	$\frac{32}{32}$	$\frac{20}{32}$	12	—
								$\frac{3}{2}$	$\frac{24}{24}$	$\frac{15}{24}$	9	—
Single-section	$\frac{64}{48}$	$\frac{180}{132}$	0.2	32	$\frac{2}{1.5}$	1	$\frac{64}{48}$	3	$\frac{32}{24}$	$\frac{20}{15}$	$\frac{12}{9}$	

Notes. voltage of heating elements, 220 V; working pressure in electric heater, up to 1 Mpa; temperature of medium in electric heater, up to 350°C.

[37, 38], but in some cases are specially designed and purchased for mounting as a nonstandardized equipment [12, 13, 36].

Of the commercially produced electric heaters, the heating units whose characteristics are presented in Table 7.10 can be recommended; they are manufactured by the Leningrad Admiralty Association [37]. A single section electric heater rated at 64 or 48 kW has the following overall dimensions: length, 2.5; width, 1; and height, 1.7 m. Increasing the number of sections entails a corresponding increase in the width of rooms allocated for the mounting of electric heaters.

Table 7.11. Characteristics of cylindrical tubular furnaces

Characteristics	Furnace type (designation)		
	TsS-1 (16/3)	TsS-1 (31/4R)	TsS-1 (31/4)
Heating surface of radiant tubes, m^2	16	31	31
Working length of radiant tubes, m	3	4	4
Heat output at average permissible heat release of radiant tubes of 100 MJ (m$^2 \cdot$ h), MJ	1600	3200	4000
Inside diameter, m	1.8	2.1	2.1
Overall dimensions with servicing platforms and stack, m: length width height	4 4 30.4	4.4 4.4 31.7	4.4 4.4 21.9
Version	1	1	11

As regards serially produced tubular furnaces, the TsS$_1$ vertical-flame cylindrical furnaces can be used, which are manufactured in two modifications: I—radiant furnaces without a convection chamber and II—radiant convective furnaces with a convection chamber (Table 7.11 [38]). In furnaces without the chamber the gases formed in the fuel combustion are from the radiant chamber directed through a conical adapter into the stack that is a continuation of the cylindrical radiant chamber of the furnace. In radiant convective furnaces the convection chamber is arranged over the radiant chamber. One or several symmetrically disposed burners, whose design depends on the fuel used, are installed in the hearth of such a furnace.

The soap-oil suspension can also be heated from 120-140°C to the temperature of formation of a homogeneous (isotropic) melt in special design electric heaters, one of which, rated at 500-550 kg/h of a grease, such as Uniol-1, is schematically shown in Fig. 7.11. Such heaters of a higher capacity can also be designed. They provide a uniform (without local overheating) and stable heating of grease. In combination with preheating of grease/semiproducts by steam, the heaters can find a wide application in manufacturing processes. They are undoubtedly attractive when fuel is in short supply.

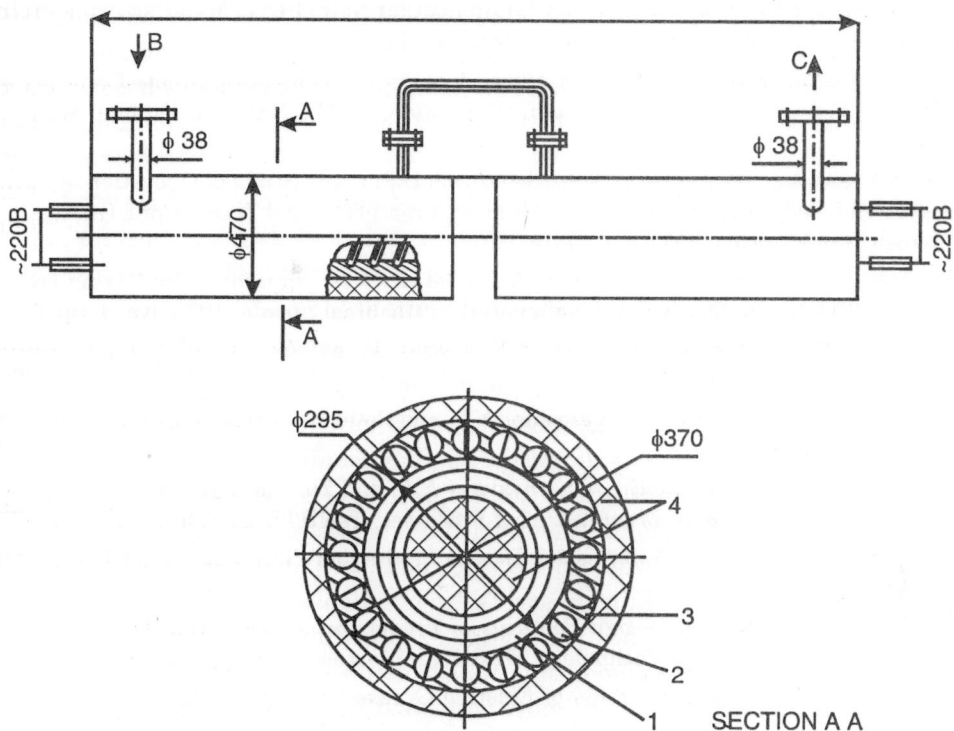

Fig. 7.11. Electric heater:
1—coil; 2—electric heating elements;
3—metal (aluminium); 4—heat insulation;
B—charging of a product; C—product discharge.

A direct "flame" heating of kettles is used sometimes in an intermittent manufacture of high temperature greases [13], but such a method of obtaining the soap melt in oil is very ineffective.

As a general rough basis for selection and evaluation of heaters it should be known that the manufacture of high temperature greases, such as lithium and calcium complex ones, involves an energy consumption of about 2000 MJ per ton of grease.

REFERENCES

1. Froishteter G.B., Trilisky K.K., Ishchuk Yu. L. and Stupak P.M., Rheologic and Thermophysical Properties of Plastic Lubricating Greases, Khimiya, Moscow, 1980.
2. Ishchuk Yu.L., Fuks I.G. and Froishteter G.B., Production of Plastic Lubricating Greases, in: Diagram Manual of Production Layouts of Petroleum and Gas Processing, Ed. by Bondarenko B.I., Khimiya, Moscow, 1983, Ch. 2, pp. 97-104.
3. Proishteter G.B., Optimum Selection of Equipment for Plastic Lubricating Grease Manufacturing Layouts, in: Plastic Greases, Abstr. of Reports at 3rd All-USSR Sci. and Techn. Conf. (Berdyansk, Sept., 1979), Naukova Dumka, Kiev, 1979, PP. 2$-27.

4. Khenkin E.M., Basic Requirements for Equipment for New Grease Manufacturing Processes, in: Improvement of Grease Manufacture, GOSINTI, 1961, pp. 52-80.
5. Froishteter G.B., Modern Processing Equipment of Plastic Lubricating Grease Factories, in: Plastic Greases, Proc. of Sci. and Techn. Conf. (Berdyansk, 1971), Naukova Dumka, Kiev, 1971, pp. 11-14
6. Nakorchevsky A.I., Skurchinsky V.A., Stretovich D.P. et al., Study of Operation of Cookers with Scraper-and-Paddle Stirrer, in: Plastic Greases, Proc. of Sci. and Techn. Conf. (Berdyansk, 1971), Naukova Dumka, Kiev, 1971, pp. 37-39.
7. Eroishteter G.B., Yurtin L.O., Ishchuk Yu. L. et al., New Apparatus for Mixing and Heat Exchange in Viscous Media, Khim. i Neftepererab. Mashinostroenie, 1975, No. 4, pp. 6-8.
8. Boner K.D., Production and. Application of Greases. Transl. from English, Ed. by Sinitsyn V.V., Gostoptekhizdat, Moscow, 1958.
9. Velikovsky D.S., Poddubny N.N., Vainshtok V.V. and Gotovkin B.D., Consistent Greases, Khimiya, Moscow, 1966.
10. Vainshtok V.V., Obelchenko E.I., Chernyavsky A.A. and Yashch-inskaya M.S., Improvement of Plastic Lubricating Grease Manufacturing Processes, TsNIITEneftekhim, Moscow, 1978.
11. Bakaleinikov M.B., Plastic Lubricating Grease Production Plant (Reminder List for Operator), Khimiya, Moscow, 1977.
12. Rudakova N. Ya., Modem Design Features in Production of Consistent Greases, in: Improvement of Grease Production Technology, GOSINTI, Moscow, 1961, pp. 32-51.
13. Garzanov G.E., Plan of Conversion of Existing Grease Production Plants and Factories, Ibid., pp. 64-80.
14. Mankovskaya N.K., Synthetic Fatty Acids, Khimiya, Moscow, 1965.
15. Tyshkevich L.V., Froishteter G.B., Ishchuk Yu.L. et al., Evaporator for Moisture Removal from Soap-Oil Suspensions, Khimiya i Tekhnologiya Topliv i Masel, 1979, N 11, pp. 34-37.
16. Ishchuk Yu. L., Kuzmichev S.P., Krasnokutskaya M.E. et al., The Present State and Future Prospects for Advances in Production and Application of Anhydrous and Complex Calcium Greases, TsNIITEneftekhim, Moscow, 1980.
17. Melnik D.T., Froishteter G.B. and Ishchuk Yu.L., Homogenation of Lithium Greases in Valve-Type Homogenizers, Khimiya i Tekhnologiya Topliv i Masel, 1974, N 1, pp. 42-46.
18. Melnik D.T. and Froishteter G.B., Study of Homogenation Conditions for Lithium and Complex Calcium Greases, in: Plastic Lubricating Greases, Additives, and Cutting Fluids, Vses. Obyed. "Neftekhim", Moscow, 1973, Issue 5, pp. 285-290.
19. Yurtin L.O., Tyshkevich L.V., Chmil V.S. and Shevchenko V.L., Experimental Study of Continuous Process of Manufacture of Lithium 12-Oxystearate Greases, in: Plastic Greases, Abstr. of Reports at 3rd All-USSR Sci. and Techn. Conf. (Berdyansk, Sept., 1979), Naukova Dumka, Kiev, 1979, pp. 106-108.
20. Technical Equipment. Homogenizers Produced by "Alfa-Laval" (FRG). Publication by "Alfa-Laval Separationstechnik" GmBH, BRD. Nr. PC 90 001 RY (in Russian).
21. Vacuum Homogenizer ENV Produced by "Koruma" (FRG). "KORUMA Vakuumenflutter EKV". VKIIPKheftekhim, No. 8022, Publication by "KORUMA Maschinenbau Paul Hauser", BRD.
22. Homogenizer Produced by "Gaulin Corporation" (USA). "GAULIN Homogenizers". VNIIPKneftekhim, No. 8033, Publication by "GAULIN Corporation", USA.

23. Yurtin L.O., Froishteter G.B., Lendyel I.V. and Shevchenko V.L. Investigation of Optimum Conditions of Deaeration of Plastic Lubricating Greases in Continuous-Action Unit, in: Plastic Greases, Proc. of 2nd All-USSR Sci. and Techn. Conf. (Berdyansk, 1973), Kaukova Dumka, Kiev, 1975, pp. 48-50.
24. Yurtin L.O., Froishteter G.B., Ishchuk Yu.L. and Zhuk E.I., Vacuum Deaerator for High-Viscosity Liquids, Information Sheet, Ser. 03-14, N 0016-74, TsNIITEneftekhim, Moscow, 1974.
25. Fully Automated Vacuum Deaeration Plant by "Koruma" (FRG). Publication by "KORUMA Maschinenbau Paul Hauser", BRD (in Russian).
26. Filters for Liquids. Catalogue by TsINTIkhinmeftemash, Moscow, 1974.
27. Chemical Processing Equipment Using Physical Methods for Intensification of Processes. Catalogue by TsINTIkhimneftemash, Moscow, 1978.
28. Logvinenko D.D. and Shelyakov O.P., Intensification of Manufacturing Processes in Units with Vortex Layer, Tekhnika, Kiev, 1976.
29. Vortex Apparatus VA-100 for Physical and Chemical Processes. Advertisement Publication, Litsenzintorg, Moscow, 1975.
30. Apparatus AVS-130V. Manufacturer^ Certificate 187.69.000.000 PS, Poltavkhimmash, Poltava, 1982.
31. Gershgal D.A. and Eridman V.M., Ultrasonic Processing Equipment, Energiya, Moscow, 1976.
32. Vodovatov JP.F., Cherevny A.A., Voiko V.P. and Libenson M,N., Lasers in Technology, Energiya, Moscow, 1975.
33. Metering Pumps and Units. Catalogue by VNIIgidromash. TsINTIkhimneftemash, Moscow, 1973.
34. Yurtin L.O., Danilevich S. Yu., .Froishteter G.B. and Kuptsov V.A., Investigation into Regularities of Efflux of Plastic Lubricating Greases from Accumulating Tanks with the Aim of Selecting Optimum Conditions at Intake of Discharging Pump, in: Chemistry, Technology, and Application of Lubricants, TsNIITEneftekhim, Moscow, 1979, pp. 63-71.
35. Vasilyeva N.I., Investigation and Development of Petroleum-Based Heat Carrier Serviceable at 300°C, Cand, Tech. Sci. Thesis, Moscow, 1978.
36. Ishchuk Yu. L., Modern State and Ways for Improvement of Manufacture of Greases at Berdyansk Experimental Petroleum-Oil Works, in: Improvement of Grease Production Technology, GOSINTI, Moscow, 1961, pp. 112-118.
37. Provisional Description, Instructions on Mounting and Operation of Section-Type Electric Heaters Operating with High-Temperature Organic Heat Carrier, Leningr. Admiral Obyed., Leningrad, 1970.
38. Tubular Furnaces. Catalogue. IsINTIkhimneftemash, Moscow, 1977.

CHAPTER 8

Automation of Manufacturing Processes. In-Process Inspection of Quality of Intermediate and of Finnished Greases

An automatic grease manufacturing process control system consists of a number of local automatic control systems for processing parameters. Such systems include an automatic control of the temperature, pressure, product level in processing equipment units, consumption of components, moisture content in the soap-oil suspension after dehydration, alkalinity in the process of saponification and quality of greases. The system can be composed either of mass produced monitoring and automation facilities of devices and instruments, or specially developed for this purpose.

Automatic measurement and control of the temperature and pressure at various stages and points of the grease manufacturing process and an automatic interlocking in response to these parameters can be effected with the use of mass produced devices of the required accuracy class [1,2]. The level in most stock tanks is automatically measured and controlled as well with the use of mass produced level gauges of different designs, operating principles and accuracy classes [1,2].

The measurement of level in apparatus with stirrers (in reactors), containing high-viscosity products (greases), is a very difficult task because of presence of moving parts as well as of grease sticking to the level gauge sensor surface. It is recommended to measure the level in such apparatus of grease manufacturing plants by hydrostatic level gauges [3], reliable in operation and simple in servicing, whose operating principle relies on measuring the pressure of the column of liquid in the apparatus on the sensing element surface.

Figure 8.1 schematically shows the design of a hydrostatic level gauge transmitter with a resilient diaphragm, whose main elements are a diaphragm seal and a pneumatic force converter. A flange with rigid corrugated diaphragm 2 built into it is attached to union 1 at the bottom of a stirrer. The space under the diaphragm is connected by a transmitting line to the receiving bellows of pneumatic-force pressure converter 3. The system is filled with a sealing liquid. The units of the transmitter are secured to the flange with the aid of a bracket. The transmitter functions as follows. The pressure in the line of the receiving bellows, produced by diaphragm 2 and depending on the height of the hydrostatic column of liquid over the latter, controls through the leverage of the pneumatic force converter the position of a flapper with respect to a nozzle. The pressure P1 in the nozzle line and hence also in the receiving chamber of pneumatic power amplifier 4 varies proportionally to the displacement of the flapper. A diaphragm unit of the pneumatic amplifier controls ball valves to vary the output pressure within 0.02-0.1 MPa.

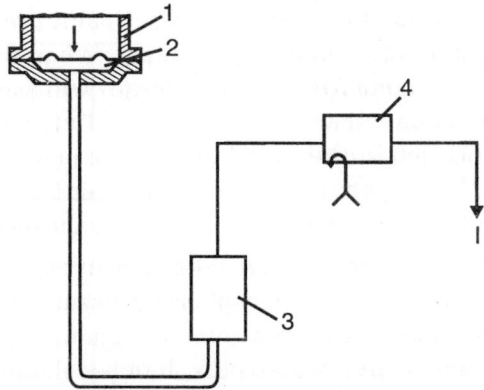

Fig. 8.1. Hydrostatic level gauge transmitter with resilient diaphragm: 1-apparatus union; 2-diaphragm; 3-pneumatic-force pressure converter; 4-power amplifier; I-pneumatic output signal.

Figure 8.2 shows the functional diagram of an automatic monitoring and signalling of the grease level in reactor-stirrer 2, based on USEPPA (Universal System of Elements of Pneumatic Productic Automatics) elements. The signal from transmitter 1 arrives to secondary indicating or recording instrument 3 and into one of the chambers of the P2ES.1 comparison

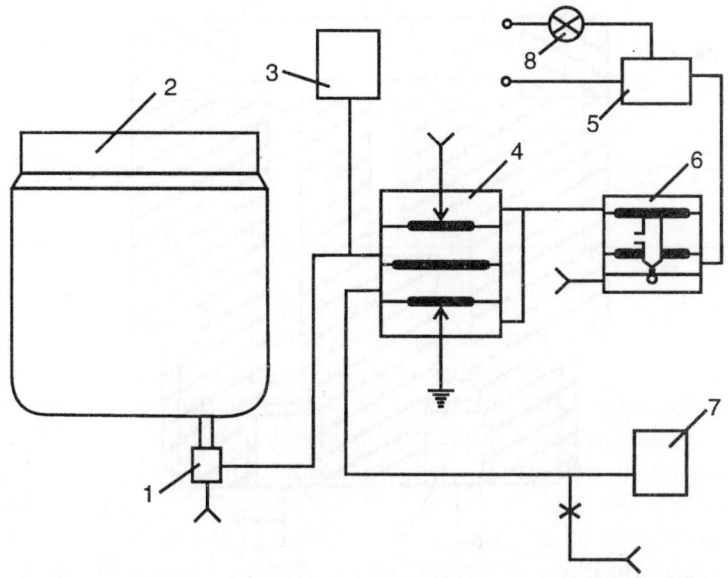

Fig. 8.2. Functional diagram of automatic monitoring and signalling of grease level in reactor-stirrer: 1-level gauge transmitter; 2-reactor; 3-recorder; 4-comparison element; 5-electropneumatic converter; 6-power amplifier; 7-set-point device; 8-signalling lamp.

element 4. The other chamber of the element receives the set-point pressure from the P2ED.3 set-point device 7. The set point can be varied over a broad range, which correspondingly changes the signalling level limit. The magnitude of the pneumatic signal from the transmitter varies with the grease level in the stirrer. When the level in the unit is the minimum, the signal at the transmitter output does not exceed, the set-point pressure, with the result that there is no signal at the output of the comparison element. When the content (grease) reaches a certain level in the stirrer, the signal from the transmitter will exceed the set-point pressure, with the result that the maximum signal will be produced at the output of the comparison element, which after amplification in the P2P.3 pneumatic power amplifier 6 will close normally open contacts of the P1PR.4 pneumo-electric converter 5 to switch on the IS signalling device 8. Any other signalling device can be used instead of the electric signalling.

The hydrostatic level gauge transmitter with a resilient diaphragm (see Fig. 8.1) is employed in apparatus operating under atmospheric pressure. For measuring the level in units operating under an excess pressure or vacuum, the pneumatic-force converter should be replaced with a differential manometer, whose plus chamber should be connected to the space under the resilient diaphragm of the transmitter and the minus chamber to the top part of the apparatus. Unfortunately, the transmitter with a resilient diaphragm cannot be used to monitor the level of high viscosity and crystallizing liquids. A force balance diaphragm transmitter of hydrostatic level gauge has been developed for this purpose 4 (Fig. 8.3). Orifice 3 in transmitter body 2 serves as a nozzle and is connected to the atmosphere, while orifice 4 is connected to the output of a pneumatic saw tooth signal generator and with the input of a control unit. Transmitter nozzle 3 is covered from above by nonresilient fluoroplastic diaphragm 1.

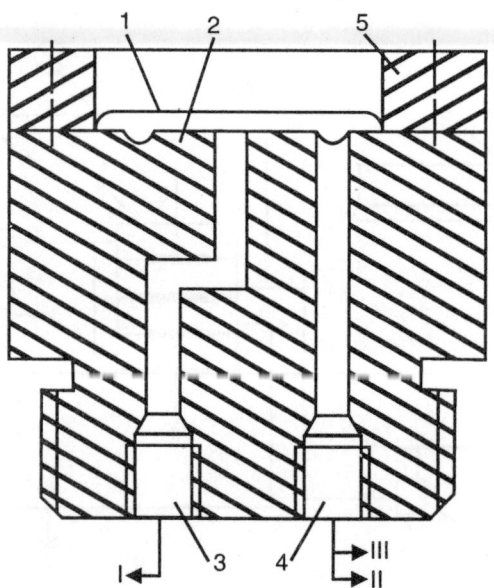

Fig. 8.3. Level gauge transmitter: 1-nonresilient diaphragm; 2-body; 3-nozzle orifice; 4-inlet orifice; 5-flange; I-to atmosphere; II-from generator; III-to control unit.

The functional diagram of measurement of the level in a stirrer with the use of the force-balance diaphragm transmitter is shown in fig. 8.4. Level gauge transmitter 1 is secured to the bottom part of the apparatus. The control unit is composed of USEPPA elements and includes maximum signal value memorizing device 2, variable throttle 3, five-diaphragm comparison element 4, set-point device 5, power amplifier 6, and pressure pulse generator 7. The transmitter functions as follows : Pulse generator 7 delivers saw-tooth pneumatic signals into the transmitter feed line under the nonresilient diaphragm. The diaphragm is acted upon from above by a pressure proportional to the liquid column height. At the minimum level in the apparatus the pressure of the liquid column on the diaphragm is low, the air fed from the generator escapes through the nozzle to the atmosphere and there is no signal at the input of the memorizing device. As the liquid level in the apparatus rises, the pressure on the diaphragm of transmitter 1 increases and the diaphragm is pressed to the nozzle so that it prevents the air escape into the atmosphere. A pneumatic signal, whose amplitude is proportional to the liquid level, appears at the control unit input. Memorizing device 2 converts a discrete signal from the transmitter into an analog signal. The memorizing time is adjusted by variable throttle 3. A doubled signal from the memorizing device is algebraically added in comparison element 4 to the signal from set-point device 5 and applied to power amplifier 6 and then to a secondary instrument, whose scale is graduated in level units. The use of pulse generator 7 upgrades the reliability of measurement, since continuous shakings of the nonresilient diaphragm prevent formation of a compacted mass in the space adjacent to the diaphragm. The frequency of the air pressure pulse is set by a variable throttle of the generator. Hydrostatic level gauges of the above described design can be used in apparatus operating under atmospheric pressure.

The UVKS-1 level gauge for highly viscous and crystallizing media 5 can also be used, whose operating principle, like that of the level gauge shown in Fig. 8.4, relies on the force-balance method of measurement of the hydrostatic pressure of the liquid column in a processing apparatus with the use of a periodically shaken nonresilient diaphragm, whose pulsation conditions are set by a pneumatic pulse generator. The UVKS-1 level gauge can be used in apparatus operating both under atmospheric and an excess pressure or vacuum. Its technical data are as follows: level measurement range, 0-10 m; viscosity of the medium under measurement, from 0.05 to 10 Pa·s; pressure in the apparatus: vacuum down to 75×10^{-4} MPa, excess pressure up to 1.6 MPa; temperature of the medium under measurement, from 10 to 220°C; and intrinsic measurement error, within ± 2.5 %.

An automatic measurement of the moisture content of the soap-oil suspension in its dewatering as well as of finished grease can be carried out using the technique for measuring the moisture content of viscous liquids, based on a full removal of moisture by heating from a thin film of grease into a stream of dry air, whose moisture content is then measured by a coloumbmetric gas moisture meter [6].

This technique has been implemented for proximate analysis of moisture in greases and additives [7,8] and in an automatic in-process grease moisture meter [9].

The automatic grease moisture meter is intended for an in process measurement and control of the moisture content of the soap-oil suspension in dehydration stage and monitoring of the moisture content of greases mainly in a continuous process of their manufacture.

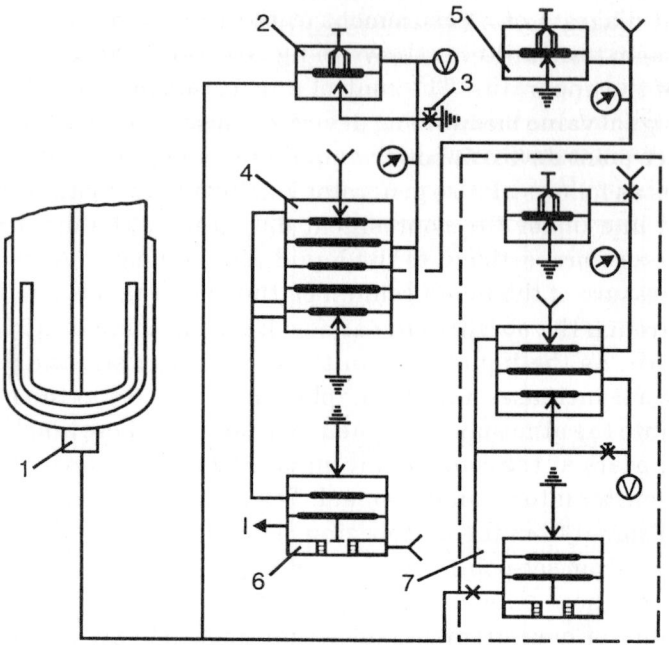

Fig. 8.4. Functional diagram of measurement of reaction mass level reactor-stirrer: 1-transmitter; 2-memorizing device; 3-variable throttle; 4-totalized; 5-set-point device; 6-power amplifier; 7-pulse generator; I-to secondary instrument.

The device for a proximate analysis of moisture is intended for a rapid check of the moisture content of the soap-oil suspension and greases mainly in intermittent manufacturing process. The functional diagram of measurement of the moisture content of greases with the aid of these devices is shown in Fig. 8.5.

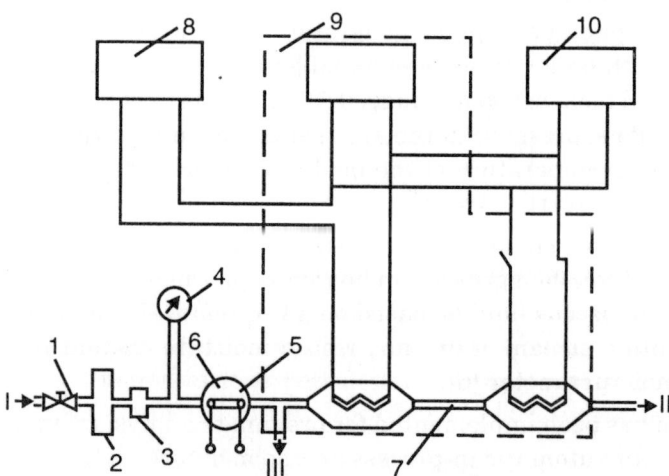

Fig. 8.5. Functional diagram of grease moisture content measurement: 1-valve; 2-drying column; 3-pressure regulator; 4-pressure gauge; 5-mass transfer-diffusion device; 6-heating element; 7-sensing element; 8-integrator; 9-coulometric gas moisture meter; 10-recorder; I-dried air; II-humid air; III-to throttling.

Dried instrumentation supply air from the factory line through valve 1 is admitted into final drying column 2 filled with dry phosphoric anhydride. The pressure of the dry air is regulated by pressure regulator 3 and monitored on pressure gauge 4. The temperature of heating element 6 of mass transfer-diffusion device 5 is maintained within 110-160°C by means of a laboratory auto transformer or a special rheostat connected at the input of a step down transformer. The moisture released from the product under test is carried over by the dry air stream into sensing element 7 of coulombmetric gas moisture meter 9, where the moisture absorbed by the air combines in the thin prosphoric anhydride film with the substance of the film to yield a concentrated solution of phosphoric acid, which is subjected to electrolysis. The latter decomposes only water and the forming oxygen and hydrogen are entrained away by the air stream

$$P_2O_5 + H_2O \rightarrow 2HPO_3 \tag{8.1}$$

$$2HPO_3 \rightarrow H_2 + \frac{1}{2} O_2 + P_2O5.$$

The electrolysis current is the measure of the moisture content of the air and hence also of the grease sample under test. Readings of the proximate moisture analysis device are taken from the scale of electrolytic current integrator 8, and of the automatic in process moisture meter, from the chart of the KSP secondary recorder [10].

The mass transfer-diffusion devices used in the proximate analysis instrument and automatic in process moisture meter differ in design. The differences allow in the former case the analysis of separate samples and in the latter case, of the product moving in a production piping with the output of a continuous signal to a secondary instrument. The design of the mass transfer-diffusion device with a metering means, employed in the proximate analyzer, is shown in Fig. 8.6.

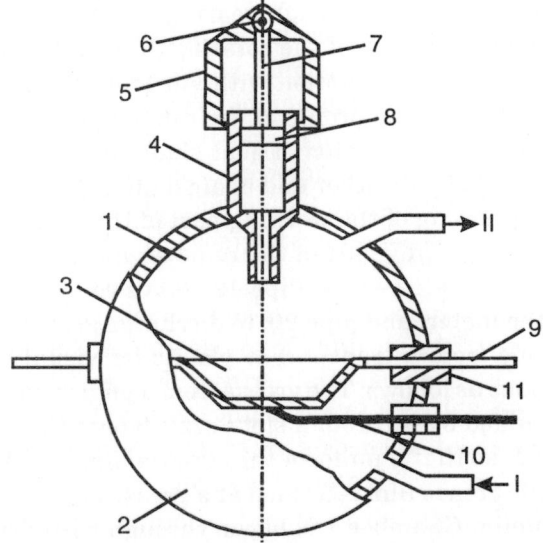

Fig. 8.6. Mass transfer-diffusion device with metering means: 1-housing; 2-cover; 3-heating element; 4-batchmeter body; 5-dial cap; 6-pin; 7-rod; 8-piston; 9-current lead; 10-thermocouple; 11-grommet; I-dry air; II-moistened air.

The mass transfer-diffusion device is a split chamber consisting of housing 1 and cover 2 and accommodating heating element 3 in the form of a dish for application of a metered amount of the product under test. Dish 5 is made from a thin tantalum plate; current-lead electrodes 9 are connected to edges of the dish, which is therefore also a heating element fed with an a.c. voltage from a step-down transformer. The temperature is monitored by chromel-alumel thermocouple 10, whose measuring junction is welded to the centre of the dish bottom, and recorded by the KSP electronic potentiometer. The current-lead electrodes are insulated from the housing by fluoroplastic grommets 11. The housing of the device has two pipe connections, for the air inlet and outlet. The metering device, consisting of body 4, dial cap 5, pin 6, rod 7 and piston 8, is secured at the top of housing 1. The body has the form of a cylinder changing over into a capillary that directs a metered amount of the grease under test onto the centre of dish 3.

Rotation of the dial cap, whose inside has a microthread, imparts a reciprocating motion to the piston, which is sealed to the body by a collar. Thus, rotation of the dial cap through a given angle forces out through the capillary a definite volume of sample from the cylinder, into which the substance under test has been taken beforehand. The analyzer provides good measurement results, especially at low moisture contents. Its technical data are as follows: measurement range, from 0.01 to 10 % water in the product; power supply 220 VAC; power consumed about 100 W; and measurement time 4-7 min.

The functional diagram of the automatic in-process grease moisture meter is shown in Fig. 8.7. Sealed chamber 1 houses disk 2 rotated from an electric motor. The disk is heated by an electric heating element mounted on the disk shaft, whose surface temperature is monitored by thermocouple 3. Union 4 serves for the dry air inlet, and union 5, for connection to coulombmetric gas moisture meter 6. The leak tightness of the chamber at the place of inlet of the disk shaft into it is provided by sealing ring 11. Metering gear pumps 7 and 8 are synchronously rotated from the electric motor through gearings. Slot nozzle 9 forms a uniform thin film of grease on the peripheral surface of the rotating disk, Ratchet gear 10, actuated from the disk shaft, imparts a reciprocating motion with vibration to special grease film remover 13, whose lug continuously follows the profile of the ratchet gear teeth. The remover base is freely put by its longitudinal slot 14 on ratchet gear shaft 15. Spring 16 is secured by its one end on the remover and thrusts by its other end against shackle 17, fixed on the ratchet gear shaft, ensuring a constant pressing of the working part of the remover to the peripheral surface of the rotating disk. The working part of the remover has at its lower end a bent tongue whose sharp edge faces grease receiver 18. Pipe 19 serves to feed grease from a production piping or apparatus into the meter, and pipe 20, to discharge grease from it. The front wall of the sealed chamber is made from a transparent plastic for visual monitoring of the meter elements. The meter functions as follows: The grease from a production piping is drawn through pipe 19 by gear pump 7 and fed by it through slot nozzle 9 onto the surface of rotating disk 2 housed in sealed chamber 1, so that a uniform thin grease film is formed on the disk surface. An 1.5 mm thick, 6 mm wide grease film is formed at a grease feed rate of 2 cm^3/min and a disk peripheral speed of 25 cm/min. Chamber 1 is blown through with dried air. In a grease residence time of 15 s on the peripheral surface of the disk at 130°C the moisture gets fully removed from it as vapor into the dry air stream, which carries the moisture over into sensing elements of coulombmetric gas moisture meter 6. The dehydrated grease film is removed from the peripheral surface of the rotating disk by special vibrational remover 13, reciprocating on the disk periphery; it smoothly moves opposite to the direction of disk rotation, but its return motion is instantaneous under the action of spring 16 at the moment when its lug, following the ratchet gear profile, comes off from the tip of a gear tooth. This provides a periodic vibration of the working part of the remover and promotes the breakaway from it of the stuck film, which then falls down into receiver 18, from which it is transferred by the gear pump through pipe 20 into the production piping. The breakaway of the stuck grease from the working part of the remover is also encouraged by a special shape and profile of the tongue at its lower end. This moisture meter offers high reliability and a good reproducibility of the results.

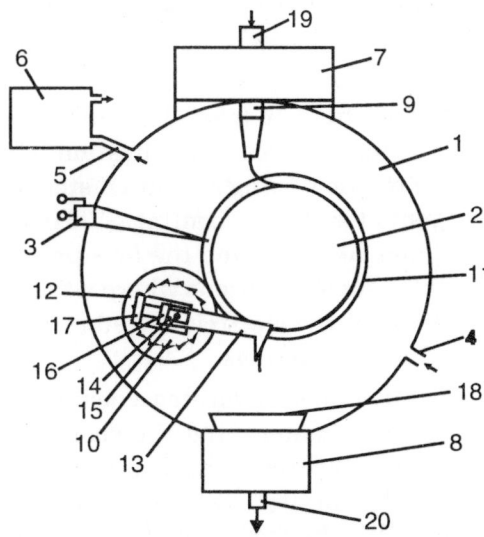

Fig. 8.7. Functional diagram of automatic in-process grease moisture meter: 1-chamber; 2-disk; 3-thermocouple; 4,5-unions; 6-coulometric gas moisture meter; 7, 8-gear pumps; 9-nozzle; 10-ratchet gear; II-ring; 12-gear; 13-grease remover; 14-slot; 15-shaft; 16-spring; 17-shackle; 18-grease receiver; 19, 20-pipes.

At a very high viscosity of a product, when it cannot be pumped by the gear pump of the automatic moisture meter, its moisture content can be measured by the DK-1 dielcometric concentratometer. This instrument, although less accurate than the automatic in-process grease moisture meter, is capable of measuring the moisture content in products even with a high content of moisture, such as in a hydrated Ca-grease, cup grease (2-3%) [6]. Its operating principle relies on variation of the permittivity of the medium under analysis with its moisture content. A great difference between the permittivity of water (e = 81) and that of the soap base (e = 2-3) ensures an accurate measurement of the moisture content of the latter. Various factors introduce, however, an error in the measurement. The instrument readings are in particular affected by variations in the flow rate, air bubbles resulting from a nonuniform operation of pumps and a high viscosity of the product, composition and concentration of the starting products, presence of surfactants, etc.

The alkali content in the soap base in the manufacture of Na-and Li-greases can be automatically measured with the aid of serially produced pH meters since a correlation exists between the alkalinity of the soap base of these greases and the pH value. An automatic system for monitoring the alkalinity of such greases differs from the traditional system for measurement of pH of liquid media as the hydraulic circuit of the reference electrode is under a pressure of 0.02 MPa. This pressure suffices to ensure the entry of the electrolyte into such a viscous medium as grease. The pressure is produced by air fed through a pressure regulator into a sealed electrolyte tank. Such an alkalinity measurement method is applicable to neutralization (saponification) processes carried out under atmospheric or smoothly varying excess pressure. A pressure regulator is used in the latter case in the reference electrode system to maintain in it a pressure equal to that of the medium under measurement. Because of this, the monitoring of soap base alkalinity with the aid of serially produced pH meters is employed in intermittent and semicontinuous processes of manufacture of Na- and Li-greases, where neutralization of acids is carried out at atmospheric pressure.

Such a method for monitoring the soap base alkalinity is unacceptable for continuous processes of manufacture of these greases since neutralization is effected there at a pulsating excess pressure. The pulsation in the flow being measured stems from operation of metering plunger pumps which feed reagents for the neutralization. There are usually at least three such pumps (in the acid, alkali, and oil lines), and the flow pressure pulsation frequency resulting from their operation creates such conditions where a follow-up pressure regulator becomes unserviceable, and this disturbs the electrolyte feed into the medium under measurement and greatly distorts the pH meter readings.

In the manufacture of calcium complex hydrated calcium and other greases there exists no correlation between the free alkali content and the soap base pH value. It is therefore expedient that an automatic control of the free alkali content in a continuous process of manufacture of these greases be effected by varying the consumption of the water-oil suspension of calcium hydroxide in accordance with the feed rate of fatty components, in particular SFA fractions or individual natural fatty acids, their acid number and activity of the slaked lime. This is carried out with the aid of a computer that computes the required feed rate of the water-oil suspension of calcium hydroxide, such as for the grease Uniol, from the well-known algorithm:

$$Q_{susp} = \frac{(Q_{SFA(h)} \cdot K_1 \, Q_{SFA(l)} \, K_2) \, 0.66}{M} \tag{8.2}$$

where Q_{sups} is the feed rate of water-oil suspension of calcium hydroxide, $Q_{SFA(h)}$ the high-molecular SFA or individual natural fatty acids (HSt, 12-HoSt), $Q_{SFA(l)}$ the feed rate of low-molecular SFA or acetic acid (HAc), K1 the acid number of the $S_{FA(h)}$, K_2 the acid number of the SFA(l), and M the calcium content in the lime used.

The functional diagram of an automatic control of the saponification stage in the manufacture of a Uniol-type cCa-grease is shown in Fig. 8.8. Special computer 1-3 continuously processes the data arriving from transmitters 1-2, 1-1 of the positions of control elements of metering pump 2 which feeds the SFA from vessel 5 and pump 5 which meters acetic acid from vessel 6 into mixer 7. The computer also receives the data from AKCh-2 acid number analyzers 1-1 and 1-2. A signal proportional to the CaO content in the lime is entered into the computer by means of manual setting device 1-4. The signal from the computer output arrives as the reference input to controller 2-3, whose second input receives a signal proportional to the current value of the feed rate of the water-oil suspension of Ca(OH)2 from transmitter 2-1 of the position of the control element of metering pump 1 that feeds the water-oil emulsion from vessel 4 for the neutralization of acids. The signal of error between the input signals of controller 2-3 acts via magnetic starter 2-4 of a multirevolution electric actuator to change the delivery of metering pump 1 in accordance with the reference input received from the computer. The delivery of metering pumps 2–3 is changed manually by rheostats 3-1, 4-1 of magnetic starters 3-2, 4-2. The computations by algorithm (8.2) can be conducted with the aid of either a computer used in the automated grease manufacturing process control system or a special microprocessor. Such an automatic system for stabilization of the soap base alkalinity can also be used in a continuous manufacture of Na- and Li-greases. In addition, a UTMK universal thermometric concentratometer can be used to measure the alkalinity in any soap based grease [10,11].

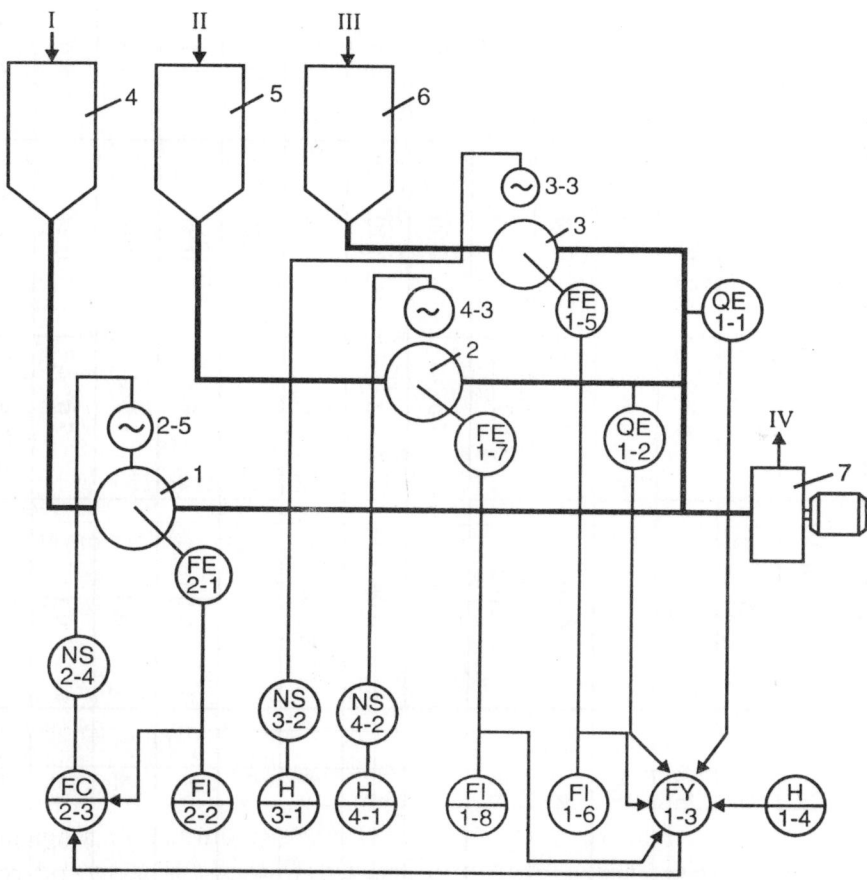

Fig. 8.8. Functional diagram of automatic control of saponification stage in manufacture of Uniol-type complex calcium greases: 1-3-metering pumps; 4-6-stock vessels; 7-mixer; I-water-oil suspension of Ca(OH)2; II-SFA; III-acetic acid; IV-to evaporator.

Studies conducted in the course of operation of a plant manufacturing the Litol-24 grease by a continuous method found a correlation between the pH value of the condensed vapour discharged from the evaporator and the reaction mixture alkalinity (12). This allowed the

Table 8.1. Results of Reaction Mixture's Alkalinity and pH of Condensate Measurements

Content of Free Alkalies in Reaction Mixture, % mass of NaOH	pH of Condensate	Content of Free Alkalies in Reaction Mixture, % mass of NaOH	pH of Condensate	Content of Free Alkalies in Reaction Mixture, % mass of NaOH	pH of Condensate	Content of Free Alkalies in Reaction Mixture, % mass of NaOH	pH of Condensate
0.10	7.5	0.70	9.0	0.26	8.1	0.00	7.0
0.08	7.7	0.46	8.7	0.44	8.2	0.16	7.5
0.28	7.8	0.24	8.2	0.40	8.5	0.18	7.8
0.20	8.4	0.20	7.8	0.26	8.4	0.28	8.0
0.34	8.5	0.16	7.7	0.54	8.8	0.58	9.0
0.42	8.7	0.08	7.5	0.56	8.9	0.52	8.5
0.56	9.0	0.06	7.2	0.10	7.5	0.30	8.4
0.48	8.8	0.02	7.0	0.02	7.2	0.20	8.0
0.24	8.0	0.62	9.0	0.00	7.0	0.04	7.0
0.14	7.6	0.38	8.6	0.02	7.1	0.06	7.3
0.08	7.4	0.30	8.2	0.10	7.4	0.12	7.9
0.00	7.1	0.12	7.6	0.24	8.0	0.34	8.3
0.04	7.3	0.04	7.3	0.60	8.5	0.02	7.2
0.28	7.7	0.04	7.1	0.48	8.9		

installation of a flow-through pH transmitter in the condensate line, where conditions for measuring the pH value turned out to be very favourable. Table 8.1 presents the results of measurements of the condensate pH value and of the alkalinity of the reaction mixture.

The correlation factor, calculated from the measurement results with the use of nonlinear correlation formulas [13], is 0.948. Such a high value evidences a correlation between the reaction mixture alkalinity and the condensate pH value. This confirms the assumption of the existence of a relation between the alkalinity of the reaction mixture and the amount of alkali getting into the condensate as a result of its mechanical carry-over with the vapour from the evaporator.

The regression equation for the reaction mixture alkalinity in the condensate pH value is

$$y = 0.068x^2 - 0.797x + 2.270 \qquad (8.3)$$

where "y" is the reaction mixture alkalinity, mass % NaOH, and "x" the condensate pH value.

The regression line for the reaction mixture alkalinity in the condensate pH value is shown in Fig. 8.9. An automatic control system for the alkalinity of the reaction mass in a plant for the manufacture of Litol-24-type greases by a continuous method, employing the configuration of a two-loop cascade system 12 is presented in Fig. 8.10.

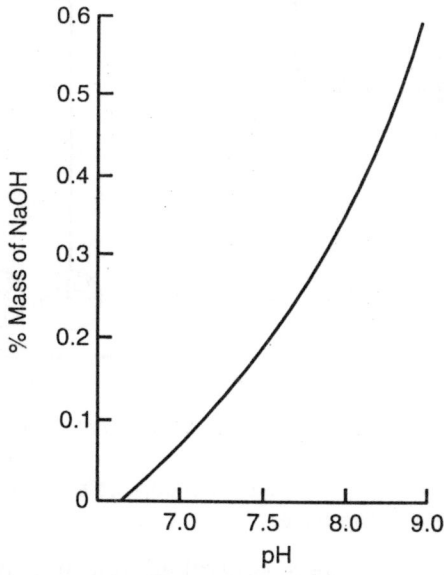

Fig. 8.9. Regression line for reaction mixture alkalinity in condensate pH value.

The secondary controller in the system is a 12-HoSt/aqueous LiOH solution flow ratio controller, acting on the flow rate of the latter. The primary controller of the cascade system is a reaction mass alkalinity controller, acting on the set-point device of the secondary controller. The secondary controller loop includes rheostat-type transmitters 3-1 and 4-1 of positions of the control elements, indicators 3-2 and 4-2 (DUP-M) of positions of the control elements, graduated, in flow rate units, normalizing converters 3-3 and 4-3 (NXP-R1M), electronic regulator 4-4 (RP2-U3), magnetic starter 4-5 (MKR-0-58), and electric actuator 4-6 (MEM-10) which varies the stroke of the plunger of the metering pump in the aqueous Li OH solution feed line.

The primary regulator loop consists of automatic thermometric concentratometer 1-1 (UTMK), secondary recorder 1-2 (KSP-3), and correcting device 1-3 (KP2-U3). The possibility for using the condensate pH value as the controlled variable is provided; in this case the primary controller loop consists of flow-through transmitter 2-1 (DM-5M) of the pH meter, high-resistance converter 2-2 (1-201.I) with a MI730A indicating instrument, recorder 2-3 (KSP2), and electronic controller 2-4 (RP2-U3) with a D-U differentiation unit.

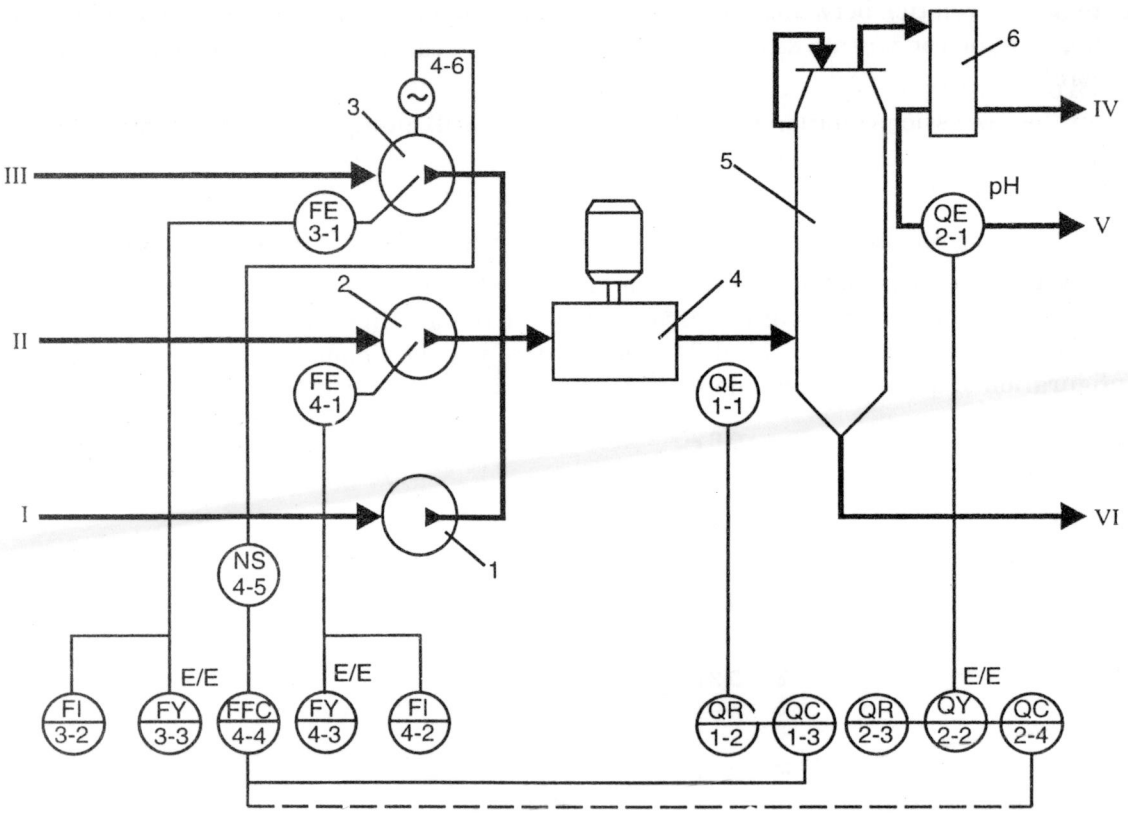

Fig. 8.10. Functional diagram of two-loop cascade automatic reaction mass alkalinity control system:
1-3-metering pumps; 4-mixer; 5-reactor; 6-condenser-cooler; I-oil; II-12-HoSt; III-LiOH; IV-vacuum; V-condensate; VI-soap base.

Important characteristics of greases, controlled in a continuous process of their manufacture, are the shear strength and viscosity. Included into these characteristics may also be the penetration, although it has no such importance as the two former parameters. A carrier of the processing information is also the velocity-viscosity characteristic (VVC), from which the correctness of processing conditions in the course of grease manufacture can be judged. The development of the procedure for an automatic in-process measurement of the grease viscosity 14 involved the task of a continuous provision of data on variation not only of the viscosity, but also of the VVC of greases. Of a wide variety of methods for measurement of the viscosity of liquids, the capillary method has been chosen for an in-process grease viscosity measurement, since viscometers based on this method offer a simple design and the highest reliability. When the capillary technique is employed, two methods of an automatic in-process monitoring of the viscosity and VVC of greases are in principle possible: the measurement of the pressure differential along a single capillary through which the grease is pumped at a varying flow rate and the measurement of the pressure differential along several series connected capillaries of different diameters, through which the grease is pumped at a constant flow rate.

The second method is more suitable for implementation, since a preset constant grease flow rate through capillaries is easy to provide by a metering gear pump with an electric speed-reducer drive.

The shear rate, in this method, is determined by the diameters and lengths of capillaries and remains a constant value for every capillary. The viscosity η by this measurement method is proportional to the pressure differential DP along the capillaries:

$$\eta_i = \frac{\tau}{D} = \frac{P_i R_i}{2LD_i} \quad (8.4)$$

where τ is the shear stress, D the average shear rate, R the capillary radius, L the capillary length, i the ordinal number of the capillary (i = 1 to 3), and K a value constant for every capillary

$$K_i = \frac{R_i}{2L_i D_i} \quad (8.5)$$

The shear rate in capillaries is adopted the same as in the grease viscosity measurement according to GOST 7163-63, i.e., of 10, 100, and 1000 s-1. The radii of capillaries are determined from the expression

$$D_i = \frac{4Q}{R_i} \quad (8.6)$$

where Q is the rate of grease flow through capillaries, determined by the expression

$$Q = G\frac{N}{Z} \quad (8.7)$$

G is the gear pump delivery per revolution, N the number of electric motor shaft revolutions per unit time and Z the speed reducer transmission ratio. In order to reduce the influence of inlet effects on the measurement accuracy, the length of capillaries is determined from the relation

$$\frac{L_i}{d_i} > 20 \quad (8.8)$$

where d_i is the capillary diameter,

It was acsertained [14] that the best reproducibility of results in parallel measurements as well as the best agreement with the data obtained on an AKV-2 instrument are attained at such an arrangement of capillaries, where the smallest diameter capillary, in which the highest of the adopted shear rates, i.e., unit , takes place, is disposed the first in the grease flow direction and diameters of the second and third capillaries increase in this direction. The reason is that the grease gets strongly deformed in the first capillary and enters the next capillary in an already broken-down state. It was noted that, when a NSh-2K3 single stage gear pump is used, the measuring system of the viscometer is sensitive to pressure variations in the production piping. Its replacement with a HSh-0,8SP3 two stage pump eliminated the pressure disturbances. Comparative measurements by the three capillary viscometer and the AKV-2 one demonstrated the best agreement of measurement results to occur when a grease sample for the AKV-2 instrument is taken downstream of the three capillary viscometer's gear pump, i.e., in a broken down state.

The use of non resilient rubber sealing diaphragms in measuring the pressure differential is little effective since the pressure transducers in this case do not return for a long time to the zero position after the load removal (gear pump stop). The same occurs also in operation of a viscometer which has volumetric seals with floating sealed metallic walls [15]. It is therefore expedient to employ volumetric seals with resilient diaphragms, capable of returning the pressure measurement system to the zero position after a load removal. Based on the above, two modifications of an automatic grease viscometer, with and without volumetric seals, have been developed and manufactured [14-16].

Figure 8.11 shows the functional diagram of an automatic in process grease viscometer with volumetric seals. Metering gear pump 1 driven from electric motor 2 through speed reducer 3 continuously pumps grease from production piping 4 through thermostatting unit 15 and three series connected capillaries 5 with different diameters, increasing in the grease flow

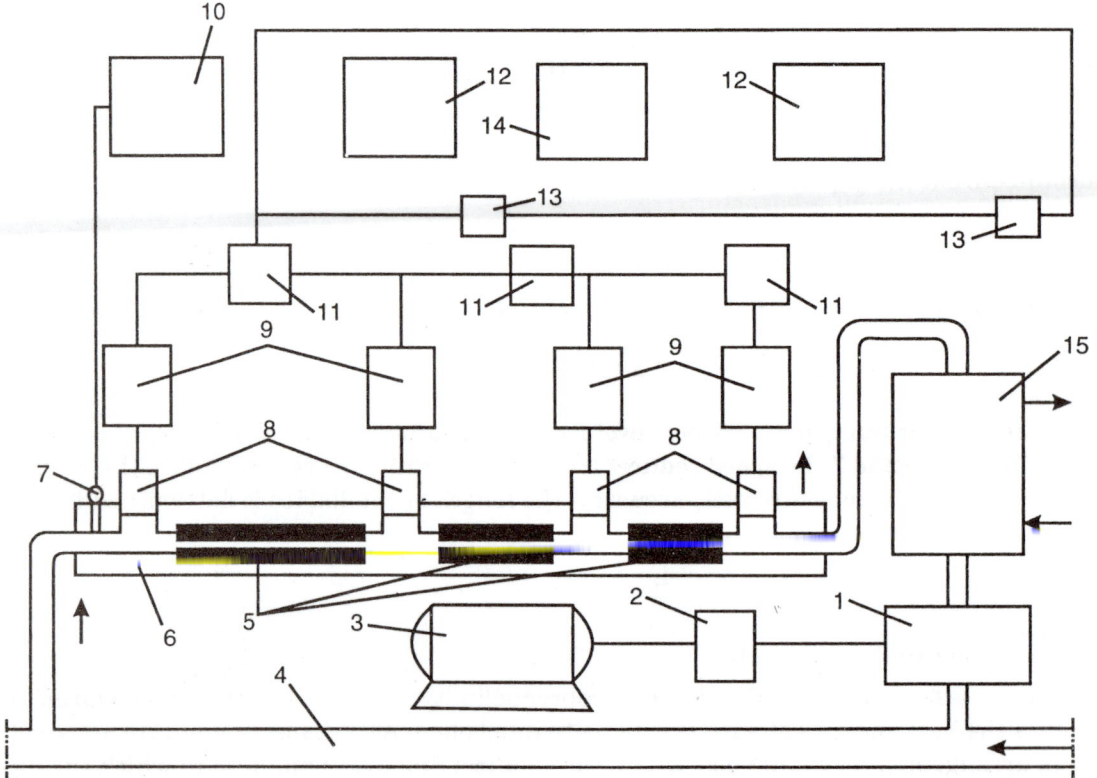

Fig. 8.11. Functional diagram of automatic in-process grease viscosimeter with volumetric seals: 1-gear pump; 2-speed reducer; 3-electric motor; 4-piping; 5-capillaries; 6-thermostatting chamber; 7-thermocouple; 8-diaphragm; 9-pressure transducer; 10-comparison element; 12-recorder; 13-multiplier-divider; 14-recorder; 15-thermostatting unit.

direction, after which the grease returns into the piping. The pressure differential along every capillary is measured with the aid of unified pressure transducers 9 (MS-P2), comparison elements 11 (P2ES3), secondary two scale recorders 12 (PV4.3E) and single scale recorder 14 (PV4.2E). The grease pressure up and downstream of the capillaries is transmitted through diaphragm sealing units 8 (PM) to unified pressure transducers 9. The diaphragm sealing units, transmitting lines and receiving bellows of unified pressure transducers are filled with a sealing liquid. The capillaries are arranged in thermostatting chamber 6. The temperature of grease at the outlet from the chamber is measured by chromel-copel thermocouple 7 connected to automatic potentiometer 10 (KSP-3). Thermostatting chamber 6, housing the capillaries, and grease thermostatting unit 15 are flown over by a heat carrier arriving from a thermostat (not shown) so that a temperature of $50 \pm 0.5°C$ is maintained. The grease flow rate through the three capillaries, differing in geometric dimensions and subjected to identical temperature conditions, being constant, the pressure difference along every capillary is, according to expression (8.4), proportional to the grease viscosity at three different fixed shear rates. The shear rate in the first (in the grease flow direction) capillary is $1000\ s^{-1}$ and in the second and third ones, 100 and $10\ s^{-1}$, respectively.

Recorder 14 and one of scales of recorders 12 are graduated in viscosity units (Pa·s).

For the VVC measurement, the pneumatic signal proportional to the viscosity at a shear rate of $10\ s^{-1}$ is in pneumatic multipliers-dividers 13 (PF1.18) divided by pneumatic signals proportional to the viscosity at shear rates of 1000 and $100\ s^{-1}$. Output signals of multipliers-dividers 13, proportional to ratios of the above viscosities, are recorded by recorders 12.

Table 8.2 presents characteristics of capillaries of the automatic grease viscometer. The grease is pumped through them by a NSh-0.8 SP3 two-stage metering pump having a rated delivery of $0.8\ cm^3/rev$ at a rotation speed of $5.83\ s^{-1}$ (35 rpm). Thus, the automatic grease viscometer continuously measures the flow at 50°C the grease viscosity at three shear rates:

Table 8.2. Characteristics of Capillaries

Number of Capillary	Shearing rate s^{-1}	Diameter mm	Length mm	Constant K
1	1000	1.7	52	8.43
2	100	3.7	110	89.45
3	10	8.0	244	881.00

$\eta_{10\ s^{-1}}^{50°C}$; $\eta_{100\ s^{-1}}^{50°C}$ and $\eta_{1000\ s^{-1}}^{50°C}$; as well as their ratio (VVC)

$$\dfrac{\eta_{10\ s^{-1}}^{50°C}}{\eta_{1000\ s^{-1}}^{50°C}}\ ;\ \dfrac{\eta_{10\ s^{-1}}^{50°C}}{\eta_{100\ s^{-1}}^{50°C}}.$$

The error of the grease viscosimeter, referred to the full scale value, is not over 8.5% with respect to the AKV-2 viscometer. The above described automatic viscometer is successfully employed in commercial production of greases by semicontinuous and continuous methods.

Figure 8.12 shows the functional diagram of an automatic in process grease viscometer without volumetric seals. The pressure differential along capillaries 5 is measured by differential manometers 1 filled with sealing liquid (glycerine, ethylene glycol or another liquid with a density higher than that of the grease, not interacting with the latter chemically and physically). Geometric dimensions of measuring capillaries are the same as those presented in Table 8.2 for the automatic viscometer with volumetric seals; characteristics of metering gear pump 8 are as well the same.

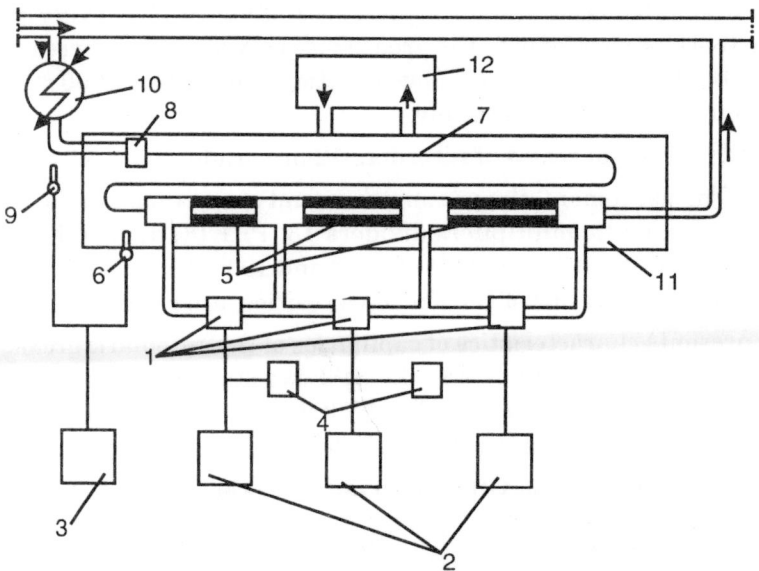

Fig. 8.12. Functional diagram of automatic in-process grease viscosimeter without volumetric seals: 1-differential manometers; 2-PV4.3E secondary recorders; 3-KSP automatic potentiometer; 4-PF 1.18 multipliersdividers; 5-capillaries; 6,9 - thermocouples; 7-coil; 8-NSh-0.8 SPZ metering gear pump; 10-heat exchanger; 11-therraostatted chamber; 12-thermostat.

Chamber 11 is thermostatted by thermostat 12. The heat carrier is an 1 : 2 mixture of oil MS-20 and transformer oil.

Thermostatted chamber 11 houses metering gear pump 8, coil 7 and capillaries 5. The thermostat maintains a temperature of 50 ± 0.5°C. When the temperature of grease in the production piping rises above 70°C, then before being fed into the instrument the grease passes through heat exchanger 10 cooled by running water. The temperature at the inlets to the thermostatted chamber and capillary units is measured by thermocouples 9 and 6 and recorded by potentiometer 3. The viscosity ratios

$$\dfrac{\eta_{10\ s^{-1}}^{50°C}}{\eta_{1000\ s^{-1}}^{50°C}} \quad \text{and} \quad \dfrac{\eta_{10\ s^{-1}}^{50°C}}{\eta_{100\ s^{-1}}^{50°C}}$$

are calculated by functional multiplier-divider 4 (PF 1.18) and recorded by secondary instruments 2 (PV4.3E). The same instruments record also the grease viscosity at 50°C at three shear rates, 10, 100, and 1000 s^{-1}. The error of the automatic grease viscometer without volumetric seals, referred to the full-scale value, is within ± 8,5 % with respect to the AKV-2 viscometer.

The automatic grease viscometer without volumetric seals is successfully employed in commercial production of greases by semicontinuous or a continuous method.

Comparing the above described functional diagrams of automatic viscometers with (see Fig. 8.11) and without (Fig. 8.12) volumetric seals, it is difficult to give preference to either of them. Viscometers without volumetric seals exhibit a higher sensitivity, but are inferior to those with volumetric seals in reliability since the sealing liquid in an unsealed system is to be periodically replenished.

When the instrument making industry starts a commercial production of miniature pressure transducers for high viscosity media, the viscometers whose functional diagram is shown in Fig. 8.11 will undoubtedly be much more attractive.

The Leningrad "Neftekhimavtomatiica" Research and Production Association has, according to the procedure and technical requirements by the UkrNDINP "MASMA", developed a VP1-U4 setup for an in process monitoring of rheological properties of greases [17, 18], which continuously measures the effective and intrinsic viscosities at three shear rates, the yield strength, and effective viscosity ratios

$$\dfrac{\eta_{10\ s^{-1}}^{50°C}}{\eta_{1000\ s^{-1}}^{50°C}} \quad \dfrac{\eta_{10\ s^{-1}}^{50°C}}{\eta_{100\ s^{-1}}^{50°C}} \quad \text{and} \quad \dfrac{\eta_{100\ s^{-1}}^{50°C}}{\eta_{1000\ s^{-1}}^{50°C}}$$

The functional diagram of the VP1-U4 setup for an in process monitoring of rheological properties of greases is shown in Fig. 8.15.

The setup consists of the following structural units: transmitter 9, which is an automatic in process grease viscometer without volumetric seals, computer 7, and data representation unit 8. The transmitter, i.e., automatic viscometer, comprises three thermostatted series-connected capillaries of different diameters, increasing in the grease flow direction. Metering gear pump 15 pumps the grease from production piping 2 through the capillaries at a constant flow rate. Pressure differentials along the capillaries are measured by differential manometers 3 whose outputs are connected to computer 7 through electropneumatic converters 6.

Used as the computer is the serially produced 15VSM-5 small computer, and as the data representation unit, the "Consul" numerical printer. The latter records the effective and the intrinsic viscosity at 10, 100, and 1000 s^{-1}.

Thermostatting unit 14 maintains the grease temperature within 50 ± 0.5 °C. The error in maintaining the grease flow rate through the capillaries is within ± 1.5% and the pressure differential measurement error, within ± 1%. The yield strength monitoring range is of 0-1200 Pa, and the viscosity monitoring range 0-600 Pa·s. The permissible intrinsic error in measuring the yield strength is ± 15% and the viscosity, ± 10%.

The VP1-U4 setup has been tested on a hydrated Ca-grease manufacturing plant at the Berdyansk PJSC "AZMOL" and is operated on a plant for manufacturing Litol-24-type Li-greases at the same works. A long time operation with these greases necessitated upgrading the reliability, which was attained by incorporation of bellows volumetric seals into the measuring system of the three-capillary viscometer.

A correlation between the effective viscosity and the penetration as well as between the effective viscosity and the shear strength has been experimentally ascertained for grease Litol-24 [21]. Table 8.3 presents the results of measurements of the effective viscosity of grease Litol-24 at 100 s^{-1} by the three-capillary viscometer and of the penetration according to GOST 5146-78. The correlation factor, calculated from these measurement results with nonlinear correlation formulas 13, is equal to 0.765.

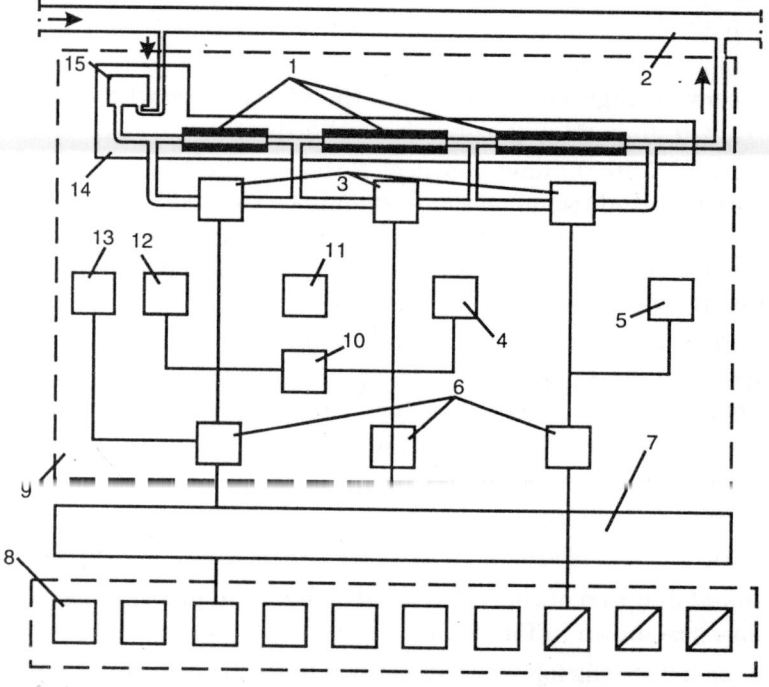

Fig. 8.13. Functional diagram of setup for in-process monitoring of rheologic properties of greases: 1-capillaries; 2-piping; 3-differential manometers; 4,5,11,12-PV4.2E secondary recorders; 6-pneumoelectric converters; 7-computer; 8-data representation unit; 9-three-capillary viscosimeter sensor; 10-PF 1.18 multiplierdivider; 13-KSP-3 automatic potentiometer; 14-thermostatting unit; 15-metering gear pump.

Table 8.3. Results of "Litol-24" Grease Viscosity and Penetration Measurement

Sample Number	Viscosity at 100 s^{-1} and 50 °C, Pa·s	Penetration, m·10^{-4}	Sample Number	Viscosity at 100 s^{-1} and 50 °C, Pa·s	Penetration m × 10^{-4}
1	184.0	197	10	119.0	208
2	70.0	296	11	119.0	218
3	287.0	198	12	135.0	233
4	58.9	256	13	58.9	259
5	250.6	217	14	93.5	243
6	126.0	267	15	78.0	255
7	45.0	306	16	37.5	251
8	109.6	225	17	93.5	240
9	135.0	210	18	93.5	247

The regression equation for the penetration in the viscosity of grease Litol-24 is

$$y = 259.66 - 0.144x - 0.004x^2,$$

where "y" is the penetration of the grease, 10^{-4} m, and x the viscosity of the grease at 100 s^{-1}, 50°C Pa·s. Figure 8.14 shows the regression line of the penetration in the viscosity of grease Litol-24. Table 8.4 presents the results of measurements of the effective viscosity at 1 s^{-1} with the aid. of the three-capillary viscometer and of the shear strength (τ_{ss}) according to GOST 7143-73 for grease Litol-24. The correlation factor, calculated from these measurement results with nonlinear correlation formulas [13] equals 0.952. The regression equation for the shear strength in the viscosity of grease Litol-24 is

$$y = 75.59 \bullet 10^{-5}x + 0.287,$$

where y is the shear strength, Pa · 10^{-2}, and x the viscosity of the grease at 1 s^{-1}, Pa · s. Figure 8.15 shows the the regression line of the shear strength in the viscosity of grease Litol-24.

The existence of such a correlation extended the capabilities of the three-capillary viscosimeter. It is used for a continuous measurement, apart from the effective viscosity and VVC, also of the penetration and shear strength. The scale of the KSP-3 automatic potentiometer 13 (Fig. 8.13) is graduated in penetration units, and of the PV4.2E secondary recorder 5, in shear strength units. Scales of secondary recorders 4 and 12 are graduated in units of viscosities at 10 and 100 s^{-1}. The effective viscosity ratio is continuously measured, by the PF1.18 multiplier-divider 10 and recorded by the PV42E secondary recorder 11.

50°C

η

10 s^{-1}

50°C

η

100 s^{-1}

Table 8.4. Results of measurements of viscosity and shear strength of grease Litol-24

Sample No.	Viscosity at 1 s⁻¹ and 50°C, Pa•c	Shear strength at 20°C, Pa
1	5200	460
2	6100	500
3	2900	220
4	1400	140
5	3000	220
6	2200	100
7	2400	220
8	2500	210

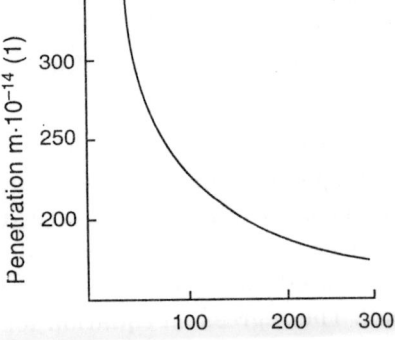

Fig. 8.14. Regression line for penetration of "Litol-24" grease in viscosity.

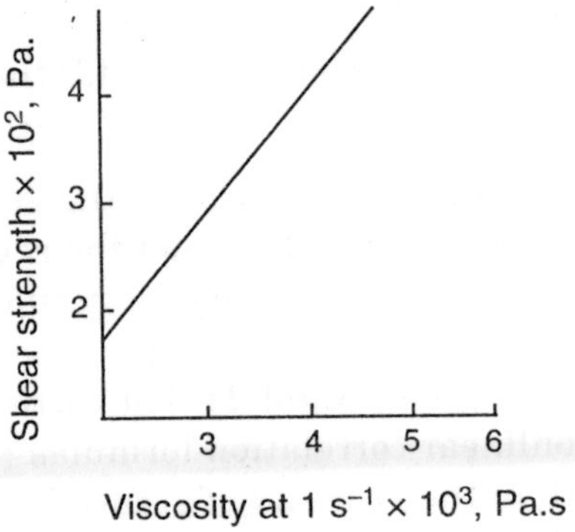

Fig. 8.15. Regression line for shear strength of "Litol-24" grease in viscosity.

It follows that three-capillary viscometer 9 is able independently (without a computer) to effect a continuous in-process measurement of the effective viscosity of a grease at three shear rates, VVC, and shear strength, and in the set with a computer can additionally calculate by the preset algorithm 19, 20 the intrinsic viscosity h_o and yield strength t_o.

Since the 15 VSM-5 computer has no output to automatic control systems, while the three-capillary viscometer can communicate with them through three pneumatic and three electric outputs, and also because the intrinsic viscosity and yield strength (in contrast to the effective viscosity) have no clear cut interpretation in the manufacturing process, the use of the computer, included in the set of the three-capillary viscometer, for monitoring and controlling commercial grease manufacturing processes becomes unnecessary. For an automatic in-process monitoring and control of the viscosity and strength characteristics of greases in commercial plants it is expedient to employ the VP three-capillary flow viscometer with resilient sealing diaphragms.

When the viscometer is used in automatic grease viscosity control systems, the pressure difference along a capillary, measured by a pneumatic differential manometer, is applied directly to a pneumatic controller which through a diaphragm type actuator controls the position of the final control element in the oil flow. When a plunger type metering pump with a variable plunger stroke is used as the final control element in the processing layout, then electro pneumatic converters are used, which convert the pneumatic output signal from the differential manometer into a standard electric signal, applied to an electronic controller whose output is connected, to a multirevolution electric actuator that varies the pump delivery.

Another method of accomplishment of the control action is employed on a plant for manufacture of grease Litol-24 by the continuous method at the Berdyansk PJSC "AZMOL". The functional diagram of such an automatic control of the grease viscosity is shown in Fig. 8.16.

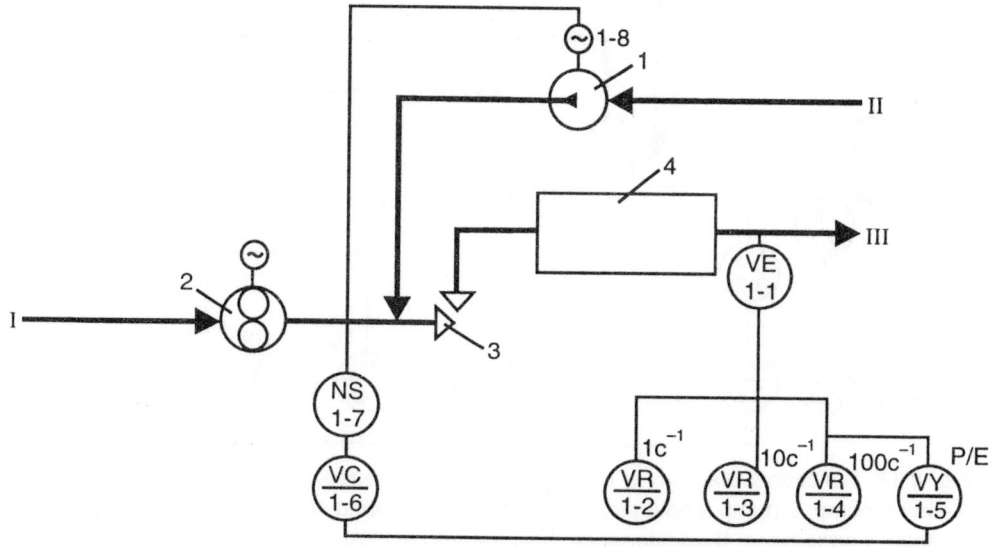

Fig. 8.16. Functional diagram of single-loop automatic grease viscosity control system: 1-metering pump; 2-pump feeding soap base from dewatering stage; 3-homogenizing valve; 4-cooler; I-soap base; II-additional oil; III-grease to accumulating tank.

The system includes automatic viscometer 1-1 (VP-1), secondary recorders 1-2, 1-3, 1-4 (PV4.2E) for three shear rates, electro pneumatic converter 1-5 (EPP-P8), controller 1-6 (RP2-U3) with a D-U differentiation unit, magnetic starter 1-7 (MKR-0-58), and actuator 1-8 (MEM-10) which varies the stroke of the plunger of the metering pump in the additional oil feed line. In operation of the system it was ascertained that an appreciable disturbing action on it is exerted by variation in the delivery of the gear pump serving as the final control element in the automatic control system for the level in the evaporator. A variation of the delivery of the pump with the level in the evaporator disturbs the ratio of flow rates of the reaction mixture and additional oil, set at a steady-state value at some time by the above-described automatic viscosity control system. Due to this, a correction for the variation in the delivery of the pump employed in the evaporator level automatic control system was incorporated into the automatic grease viscosity control system, which resulted in a two-loop cascade automatic grease viscosity control system, whose functional diagram is shown in Fig. 8.17.

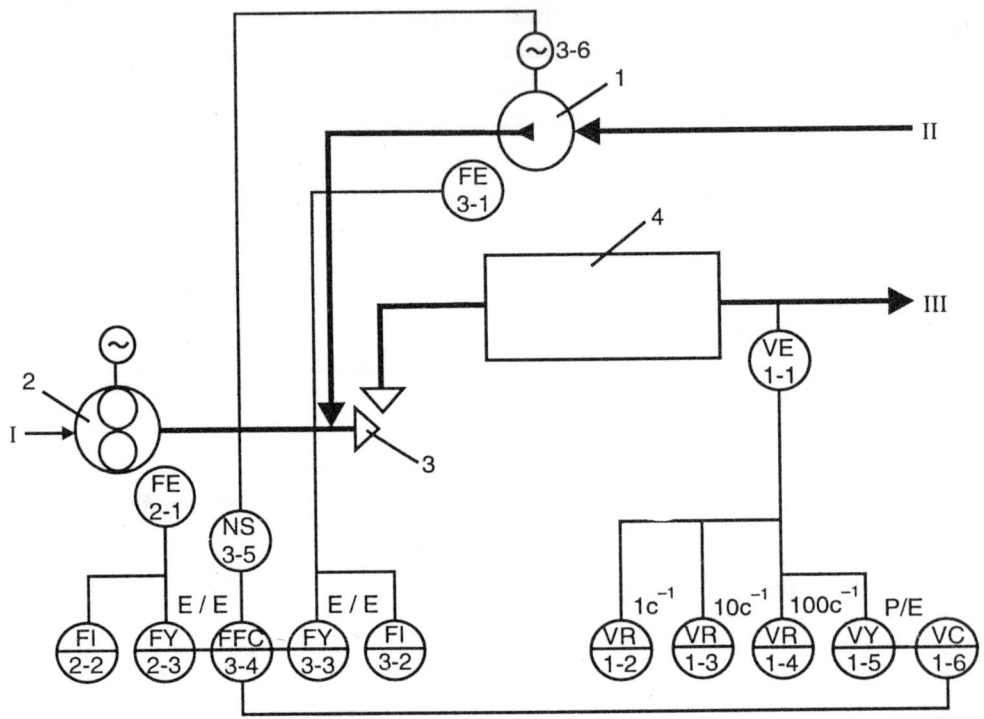

Fig. 8.17. Functional diagram of two-loop automatic grease viscosity control systems: I-metering pump; 2-pump feeding soap base from dewatering stage; 3-homogenizing valve; 4-cooler; I-soap base; II-additional oil; III-grease to accumulating tank.

The secondary controller in the system is a soap base from the evaporator/additional oil flow ratio controller, acting on the flow rate of the latter. The primary controller of the cascade system is a grease viscosity regulator, acting on the set-point device of the secondary controller. The secondary controller loop includes rheostat type transmitters 2-1, 3-1 of positions of final control elements, remote indicators 2-2, 3-2 (DUP-M) of positions of the control elements, graduated in flow rate units, normalizing converters 2-2, 3-3 (NP-R1M), controller 3-4 (RP2-U3), magnetic starter 3-5 (MKR-0-58), and actuator 3-6 (MEM-10B) that varies the stroke of the plunger of the metering pump in the additional oil feed line.

The primary controller loop consists of automatic viscometer 1-1 (VP-1), secondary recorders 1-2, 1-3, and 1-4 (PV4.2E), electro pneumatic converter 1-5 (EPP-P8), and corrector 1-6 (KP2-U3) with a D-U differentiation unit.

The VP-1 viscometer is also employed in the automatic shear strength control system for grease Litol-24 on the same plant; its functional diagram is shown in Fig. 8.18.

Apart from the VP-1 viscometer 1-1, the shear strength control system includes secondary recorder 1-2 (PV10.1E) graduated in shear strength units, pneumatic controller 1-3 (PR3.21), logic conversion unit 1-4, which has an output to pneumatic actuator 1-5 installed in the coolant line to the cooler and through electro pneumatic converter 1-6 (EPP-P8), to controller 1-7 (RP2-U3) with a D-U differentiation unit, magnetic starter 1-8 (MKR-0-58) and actuator 1-9 (MEM-10B) that varies the stroke of the plunger of the metering unit which feeds 12-HoSt for the soap base preparation.

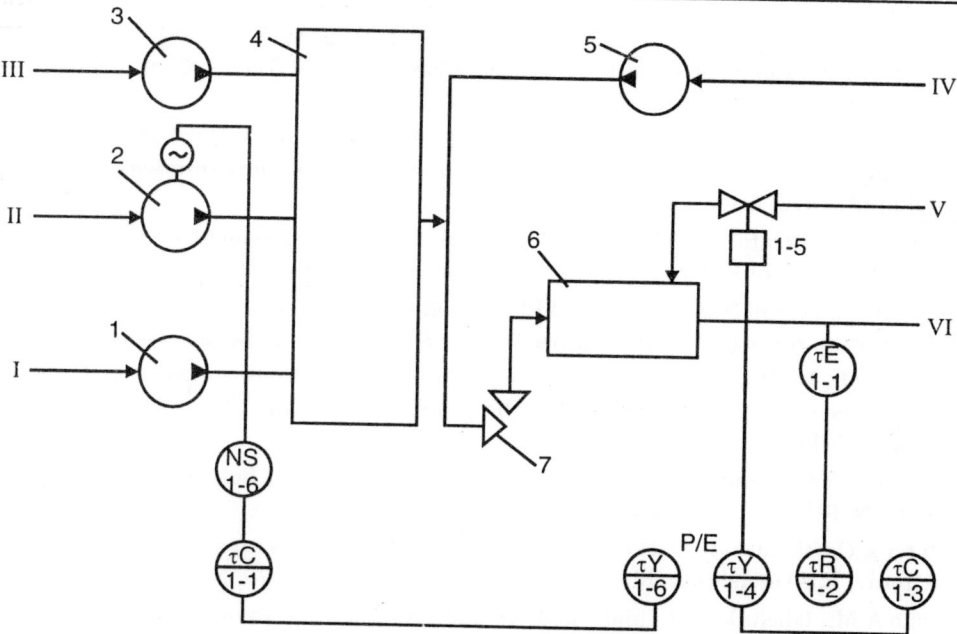

Fig. 8.18. Functional diagram of automatic shear strength control system for Litol-24-type greases: 1, 2, 3, 5-metering pumps; 4-reactor-saponifier; 6-cooler; 7-homogenizing valve; I-oil; II-12-HoSt; III-LiOH; IV-additional oil; V-coolant; VI-grease to accumulating tank.

The stabilization of the penetration and shear strength at the specified level with the aid of the above described automatic control systems provides for the required level of other quality characteristics of grease Litol-24.

Such a viscometer can be employed for an automatic control of quality of other greases in their manufacture. It should be remembered, however, that the correlation, between the viscosity and other rheological characteristics varies with the grease type and is to be experimentally determined for graduation of scales of the secondary instruments recording the characteristics, while the graduation of secondary instruments in the effective viscosity remains unchanged for all grease types.

REFERENCES

1. Automatic Instruments, Controllers, and Computing Systems. Reference Handbook, Ed. by Kosharsky B.D., Mashinostroenie, Leningrad, 1976.
2. Rudakova N. Ya., Modern Design Features in Production of Consistent Greases, in: Improvement of Grease Production Technology, GOSINTI, Moscow, 1961, pp. 32-51.
3. Manoilo A.M., Chernov N.G. and Latash V.N., Measurement of Grease Level in Apparatus with Stirrers, in: Plastic Greases, Proc. of 2nd All-USSR Sci. and Techn. Conf. (Berdyansk, 1975), Naukova Dumka, Kiev, 1975, pp. 147-149.
4. Inventor's Certificate 444, 945 (USSR). Apparatus for Liquid Level Measurement. A.M.Manoilo, L.V.Kruzement-Prichodko. Publ. in B.I., 1974, N 36.

5. Zhuravlev L.P., Gostev B.I. and Morosov Yu. A., Level Gauge for Viscous Crystallizing Media, in: Plastic Greases, Abstr. of Reports at 3rd All-USSR Sci. and Techn. Conf. (Berdyansk, Sept., 1979), Naukova Dumka, Kiev, 1979, pp. 144-146.
6. Manoilo A.M., Terebii N.M., Ishchuk Yu.L. et al., Automated Methods of Measurement of Moisture Content of Greases and Additives, Neftepererabotka i Neftekhimiya, Vsesoyus. Obyed. "Neftekhim", 1976, Issue 11, pp. 28-36.
7. Inventor's Certificate 372,489 (USSR). Apparatus for Measurement of Moisture Content in Greases. Yu.L.Ishchuk, A.M.Manoilo, V.G.Primachenko et al. Publ. in B.I., 1973, N 13.
8. Manoilo A.M., Terebii N.M. and Korbut I.R., Semiautomatic Meter of Moisture Content of Greases and Additives, Khimiya i Tekhnologiya Topliv i Masel, 1975, N 11, pp. 51-53.
9. Manoilo A.M., Terebii N.M. and Abushek V.A., Automatic In-Process Grease Moisture Meter, in: Plastic Greases, Proc. of 2nd Sci. and Techn. Conf. (Berdyansk, 1975), Naukova Dumka, Kiev, 1975, pp. 145-146.
10. Chepchurov Ya.I., Universal Thermometric Concentratometer, Belgorod, 1983 (Inform. Sheet, MNTsTI, N 207).
11. Inventor's Certificate 407, 217 (USSR). Thermometric Goncentratometer. Ya.I.Chepchurov, E.L.Ozerov, R.B.Goldshtein, M.N.Bok-shitsky. Publ. in B.I., 1973, N 46.
12. Manoilo A.M., Ishchuk Yu.L. and Rudovich I.M., Automatic Measurement and Control of Soap Base Alkalinity in Manufacture of Plastic Lubricating Greases, Khim. Tekhnologiya, 1985, N 1, pp. 57-59.
13. Batuner L.M. and Posin M.E., Mathematical Methods in Chemical Engineering, Khimiya, Leningrad, 1971.
14. Manoilo A.M., Automatic In-Process Measurement of Effective Viscosity of Plastic Lubricating Greases, Neftepererabotka i Nef-tekhimiya, Kiev, 1982, Issue 22, pp. 26-32.
15. Manoilo A.M., Froishteter G.B., Latash V.N. et al., Automatic Instrument for Continuous In-Process Measurement of Effective Viscosity of Greases, Ibid., 1976, Issue 14, pp. 55-58.
16. Manoilo A.M., Froishteter G.B., Latash V.N. et al., Commercial Tests of Instrument for Continuous In-Process Measurement of Effective Viscosity of Greases, in: Plastic Greases, Proc. of 2nd All-Union Sci. and Techn. Conf. (Berdyansk, 1975), Naukova Dumka, Kiev, 1975, pp. 146-147.
17. Inventor's Certificate 873,035 (USSR). Method for Continuous Determination of Rheologic Properties of Plastic Disperse Systems. G.B. J Troish.teter, A.M.Manoilo, K.K. Trilisky et al. Publ. in B.I., 1981, N 38.
18. Froishteter G.B., Manoilo A.M., Trilisky K.K. et al., Instrument for Automatic In-Process Monitoring of Rheologic Properties of Plastic Lubricating Greases, in: Plastic Greases, Abstr. of Reports at 3rd All-USSR Sci. and Techn. Conf. (Berdyansk, Sept., 1979), Naukova Dumka, Kiev, 1979, pp. 142-144.
19. Froishteter G.B., Trilisky K.K., Manoilo A.M. et al., Method for In-Process Measurement of Rheologic Properties of Plastic Lubricating Greases, in: Plastic Greases, Proc. of 2nd All-USSR Sci. and Techn. Conf. (Berdyansk, 1975), Naukova Dumka, Kiev, 1975, pp. 164-166.
20. Froishteter G.B., Manoilo A.M. and Trilisky K.K., Method for In-Process Measurement of Rheologic Properties of Plastic Disperse Systems, in: Abstr. of Reports at Workshop "Instrumental Methods of Rheology", Sept. 1977, VDKKh SSSR, Moscow, 1977, p. 17.
21. Manoilo A.M., Ishchuk Yu. L., Rudovich I.M. et al., Automatic Control of Viscous and Strength Characteristic of Plastic Lubricating Greases in Process of Their Manufacture, Khim. Tekhnologiya, 1985, N 2, pp. 48-50.